A Brief History of Geology

Geology as a science has a fascinating and controversial history. Kieran D. O'Hara's book provides a brief and accessible account of the major events in the history of geology over the last 200 years – from early theories of Earth's structure during the Reformation, through major controversies over the age of the Earth during the Industrial Revolution, to the more recent twentieth-century development of plate tectonic theory, and on to current ideas concerning the Anthropocene. Most chapters include a short feature box providing more technical and detailed elaborations on selected topics. The book also includes a history of the geology of the Moon, a topic not normally included in books on the history of geology. The book will appeal to students of Earth science, researchers in geology who wish to learn more about the history of their subject, and general readers interested in the history of science.

KIERAN D. O'HARA is Professor Emeritus in the Department of Earth and Environmental Sciences at the University of Kentucky. He has published more than 40 articles in international journals and has received numerous research awards from the American National Science Foundation. He taught geology at undergraduate and graduate levels at the University of Kentucky for 30 years. His other books include *Cave Art and Climate Change* (2014) and *Earth Resources and Environmental Impacts* (2014).

'O'Hara does a great job of covering both the old (late 1800s) and the new (1960-1970) history of geology. Included are informative, but concise, biographies of all the major players in the nineteenth and twentieth centuries. The author shows very clearly how Wegner's continental drift – which was not originally accepted by the scientific community – came together with Harry Hess's seafloor spreading in the 1970s, and led to the "Great Plate Tectonic Revolution" in the Earth sciences. I really liked the chapter on isotopic dating, where the author clearly explains how geologists learned to use isotopes to date geologic events – no other book on the history of geology illustrates this so clearly.'

– Kent Condie, New Mexico Institute of Mining and Technology

'The nearly four-century existence of geology as a concept – "the study of the Earth with its Furniture" as it was first put – has been mired in periods of uncertainty, revolution, speculation and controversy. O'Hara has tied it all up in a concise, neatly arranged and highly readable summary, essential to all who want to know more of the fascinating story of this most fundamental of sciences.'

– Simon Winchester, author of *The Map That Changed the World*

A Brief History of Geology

KIERAN D. O'HARA
University of Kentucky

CAMBRIDGE
UNIVERSITY PRESS

Shaftesbury Road, Cambridge CB2 8EA, United Kingdom

One Liberty Plaza, 20th Floor, New York, NY 10006, USA

477 Williamstown Road, Port Melbourne, VIC 3207, Australia

314–321, 3rd Floor, Plot 3, Splendor Forum, Jasola District Centre, New Delhi – 110025, India

103 Penang Road, #05–06/07, Visioncrest Commercial, Singapore 238467

Cambridge University Press is part of Cambridge University Press & Assessment, a department of the University of Cambridge.

We share the University's mission to contribute to society through the pursuit of education, learning and research at the highest international levels of excellence.

www.cambridge.org
Information on this title: www.cambridge.org/9781107176188

DOI: 10.1017/9781316809990

First published 2018

A catalogue record for this publication is available from the British Library

Library of Congress Cataloging-in-Publication data
Names: O'Hara, Kieran D., author.
Title: A brief history of geology / by Kieran D. O'Hara.
Description: Cambridge : Cambridge University Press, 2018. |
    Includes bibliographical references and index.
Identifiers: LCCN 2017051822 | ISBN 9781107176188 (hardback : alk. paper) |
    ISBN 9781316628294 (pbk. : alk. paper)
Subjects: LCSH: Geology–History. | Geologists–Biography.
Classification: LCC QE11 .O33 2018 | DDC 551.09–dc23
    LC record available at https://lccn.loc.gov/2017051822

ISBN   978-1-107-17618-8   Hardback
ISBN   978-1-316-62829-4   Paperback

Contents

Preface

Charles Lyell's *Principles of Geology* (Vol. 1, 1830) is one of the earlier treatments of the history of geology, sketching progress throughout the history of geology. In that history, Lyell takes the Neptunist school to task for believing that all rock successions, including igneous rocks, were precipitations from a "chaotic fluid." He also praised the English surveyor, William Smith, who by 1790 had recognized that stratified rock formations could be identified by their fossil assemblages and who, by himself, produced the first geologic map of England and Wales in 1815 – this work is recounted in Simon Winchester's book *The Map That Changed the World* (2001). Subsequent histories of geology include Von Zittel's *History of Geology and Paleontology* (1901), Archibald Geike's *The Founders of Geology* (1905), and *The Birth and Development of the Geological Sciences* by Frank Adams (1938), which includes a chapter on ancient Greek and Roman writers.

 Geology in the Nineteenth Century by M. T. Greene (1982) focuses on tectonic theories, the origin of mountain belts, and Continental Drift. The *Rejection of Continental Drift* by Naomi Oreskes (1999) discusses the major players involved in the arguments on both sides of the Atlantic on this issue. *The Ocean of Truth* (1986) is a personal account by H. W. Menard of discoveries made in the oceans in the 1950s and 1960s leading up to, but prior to, the discovery of plate tectonics. A compilation of historically important academic papers on plate tectonics was produced by Alan Cox entitled *Plate Tectonics and Geomagnetic Reversals* (1973). The introduction by Cox sheds substantial light on the history of plate tectonics. This compilation also includes a 1944 paper by the British geologist Arthur Holmes, whose contribution to global tectonics has largely been underestimated.

A series by H. R. Frankel entitled *The Continental Drift Controversy* was published in 2012 by Cambridge University Press and included four volumes: Volume 1, *Wegener and the Early Debate;* Volume 2, *Paleomagnetism and Confirmation of Drift;* Volume 3, *Introduction of Seafloor Spreading;* and Volume 4, *Evolution into Plate Tectonics.* Three important books by the historian and geologist Martin Rudwick published by Chicago University Press, are *Earth's Deep History* (2014), *Worlds before Adam* (2008), and the *Great Devonian Controversy* (1985). *Great Geologic Controversies* by A. Hallam (1989) is also notable (Oxford University Press). *Two Hundred Years of Geology in America,* edited by C. J. Schneer (University Press of New England, 1979) confines itself to American geology. *History of Geology* by H. B. Woodward (Arno Press, 1978) confines itself to the seventeenth and eighteenth centuries. *A History of Geology* by G. Gohau (Rutgers University Press, 1991), translated from the French, takes a somewhat Eurocentric view of the history of geology.

The present book begins in the 1780s with James Hutton and other founders of geology, and ends with the current proposal for an Anthropocene epoch, based on environmental considerations. I conclude the Anthropocene should be an archaeological subdivision rather than a geological one. Other topics are Continental Drift (Chapter 6) and the Pleistocene Ice Age (Chapter 9). Two topics not normally treated in a history of geology but included in this book are the history and origin of the Moon (Chapter 10) and a brief history of isotope geology (Chapter 8). A crisis in the subdiscipline of tectonics is identified in Chapter 5, beginning circa 1890s, and this crisis does not end until the discovery of plate tectonics in the 1960s (Chapter 7). The origin of igneous rocks is covered in Chapter 4. The final chapter attempts to examine whether the history of geology (not just that of plate tectonics) fits into the Kuhnian scientific revolution framework; I conclude that it does. Geologic jargon is kept to a minimum in the hope that the book will be of interest not just to Earth science students and teachers, but also to a wider general audience. I would like to thank those who reviewed various chapters of the book: Steven Greb, Frank Ettensohn, David Moecher, Gustave Lester, Kent Ratajeski, Sean Bemis, and Malcom Rutherford.

I Major Nineteenth-Century Players

From the standpoint of Catastrophism little progress was made.
Uniformity proved a great advance, but in detail it is apt to lead us astray
if applied too dogmatically.

– Arthur Holmes, 1913[1]

INTRODUCTION

The most important book on geology published in the nineteenth century was probably *Principles of Geology* by Charles Lyell (in three volumes, 1880–1883), which instigated one of the major scientific debates that raged throughout the nineteenth century – namely that of the catastrophists versus the uniformitarians.[2] The dichotomy set up by these two groups first appeared in a review of volume two of Lyell's *Principles* in 1832.[3] The subtitle to the first edition of the *Principles* elaborates one of the book's main goals: "Being an attempt to explain the former changes of the Earth's surface by reference to causes now in operation." This statement assumed that the physical laws operating today also operated in the past, consistent with the immutability of the laws-of-nature idea – an idea accepted by most philosophers at that time, with the exception of some biblical literalists or scriptural geologists who entertained preternatural causes.[4]

Throughout his book, however, Lyell indicates that not only were the kinds of processes in the past the same as today, but in addition their *intensity* was also the same (see Box 1.1). On this, Lyell received a lot of hostile opposition, especially from geologists who saw evidence in the geologic record of "revolutions" – namely species extinctions, inundations and recessions of the seas indicated by sharp changes in the fossil content of strata, faults that juxtaposed contorted strata with horizontal strata, and mountain-building

BOX I.I   **Extracts from Lyell's Letters**[22]

Italics are original. Explanatory notes in square brackets by author.

To Murchison   Naples: Jan. 15, 1829

My dear Murchison, ... I will tell you fairly that it is at present of no small consequence to me to get a respectable sum for my volume – not only to cover extra expenses for present and future projected campaigns ... My work is in part written, and all planned ... it will endeavour to establish the *principle of reasoning* in the science; and all my geology will come in as illustration of my views of those principles, and as evidence strengthening the system necessarily arising out of the admission of such principles, which, as you know, are neither more or less than that *no causes whatever* have from the earliest time to which we can look back, to the present, ever acted, but those *now acting;* and that they never acted with different degrees of energy from that which they now exert. I must go to Germany and learn German geology and the language ... If I can but earn the wherewith to carry on the war, or rather its *extraordinary* costs, depend upon it I will waste no time in bookmaking for lucre's sake.

To His Sister   Rome: Jan. 21, 1829

My dear Marianne, ... Longman [a publisher] has paid down 500 guineas [roughly equivalent to 500 pounds sterling] to Mr. Ure of Dublin for a popular work on geology, just coming out [*A New System of Geology*]. It is to prove the Hebrew cosmogony, and that we ought all to be burnt in Smithfield [a site in London used for execution of heretics in earlier times]. So much the better. I have got a rod for the fanatics, from a quarter where they expect it not. The last Pope did positively dare to convoke a congregation and *reverse* all that his predecessors had done against Galileo, and there was only a minority of one against. How these things are so little known in Paris and London, heaven knows.

To Dr. Fleming   June 10, 1829

My dear Sir - ... Buckland was so amazingly annoyed at my having such an anti-diluvialist paper read [at the Geological Society], that he got Conybeare to write a controversial essay on the Valley of the Thames, in which he drew a comparison between the theory of the Fluvialists, as he terms us, and the Diluvialists, as (God be praised) they call themselves. ... But you must know that Buckland and Conybeare, distinctly admit three universal deluges, and many catastrophes, as they call them, besides! But more of this when we meet.

events.[5,6] These observations suggested that nature was not uniform in its intensity in the geologic past. Geologists who saw the importance of revolutions and intense, rapid transformations in the rock record came to be known as catastrophists (a term that was probably regarded as an overstatement by many experts of the time). Catastrophists saw that the geologic record was not uniform or cyclic as Lyell and his predecessors Hutton and Playfair had argued.[7] Lyell insisted that, on average, internal processes (e.g., earthquakes and volcanoes) and surficial processes (e.g., rivers, tidal currents, and climate) were of the same intensity globally in the geologic past as they are today.[2] He also applied this uniformity principle to the organic realm, and he rejected the French botanist Jean Lamarck's (1774–1829) proposed theory of biological inheritance, also known as transformation of the species. Lyell's seminal book is essentially a summary of all the known facts about geological processes that operated on the surface of the Earth throughout recorded human history (the past few thousand years), and asserts that these processes alone are sufficient to explain the past geologic record going back millions of years; catastrophic events or revolutions were not required. Remarkably, with the exception of parts of the third volume, there is very little actual geology in *Principles*. Its main emphasis is on the historical record.

The second controversy at this time was that of the Neptunists (whose chief proponent was Abraham Werner, together with his students) and that of the Plutonists (whose chief proponent was James Hutton, (popularized by John Playfair). The Neptunists thought all rocks were precipitated from a global ocean, including igneous rocks. The Plutonists, on the other hand, recognized igneous rocks for what they were, namely, derived from magma. The two controversies are somewhat intertwined: uniformitarians were generally Plutonists, and catastrophists were generally Neptunists. That Neptunists were also catastrophists is hardly surprising since they required a global menstruum, or primeval ocean from which all rocks were precipitated, and this ocean advanced and receded globally more than once

Table 1.1 *Principal nineteenth-century players*

Name	Lifespan	Training	U/C	Age in 1830*
Hutton, James	1726–1797	Medicine	U	–
Werner, Abraham	1749–1817	Mining/mineralogy	C	–
Cuvier, Georges	1769–1832	Natural history	C	61
Smith, William	1769–1839	Surveyor	?	61
Buckland, William	1784–1856	Theology	C	46
Sedgwick, Adam	1785–1873	Theology/math	C	45
Conybeare, William	1787–1857	Theology	C	43
Murchison, Roderick	1792–1871	Military	C	38
Lyell, Charles	1797–1875	Law	U	33
Agassiz, Louis	1807–1873	Medicine	C	23
Darwin, Charles	1809–1882	Medicine/theology	U	21

Note: U: uniformitarian; C: catastrophist; * the year *Principles* was first published.

during major Earth revolutions. That uniformitarians were generally also Plutonists is something of a historical accident reflecting the fact that Hutton saw igneous activity as causing rejuvenation of the landscape after being denuded by uniform erosional processes operating today.[8] The Neptunism–Plutonism controversy is addressed in more detail in Chapter 4 in the context of the origin of igneous rocks.

The purpose of this chapter is to introduce the reader to the main players active in these debates by providing a brief biographical sketch for each author followed by some comments on their historical role; some of these authors will be encountered again in subsequent chapters. Table 1.1 summarizes the players chronologically, in order of their date of birth. Due to space constraints, the list is highly selective.

Their educational training and their age at the time *Principles* was first published is also shown, in order to provide some historical context. Whether they were uniformitarians or catastrophists is also noted. The nineteenth century was one of the most scientifically active centuries for the nascent disciplines of geology and paleontology. The first three authors (Hutton, Werner, and Cuvier) largely belong to the late eighteenth century, but their views had a strong influence on nineteenth-century debates and remain important to this day.

James Hutton (1726–1797). James Hutton was born in Edinburgh, and his father, a wealthy merchant, died when he was three.[9] He inherited property and enough wealth that he did not have to earn a living. He entered the University of Edinburgh in 1740 to study the humanities; he reentered the university again in 1744 to study medicine. He spent two years in Paris beginning in 1747, where he developed his interest in chemistry and geology. He received a medical degree from the University of Leiden in the Netherlands in 1749, but he never practiced medicine. In 1750 he retired to Edinburgh where he took up farming on his inherited property southeast of the city. He also engaged with a friend in a successful business involving the manufacture of ammonium chloride (referred to as *sal ammoniac* at the time), which was used in industrial processes and probably added to his wealth. In 1754 he traveled widely in northern Europe to study farming methods, and also developed his increasing interest in geology. After fourteen years of farming, he moved to Edinburgh in 1768 where it appears he undertook experiments in chemistry and also collected fossils. (At the time the word fossil included both minerals and organic remains.) He became an active member in what would become the Royal Society of Edinburgh in 1783. He was friends with members of the Scottish Enlightenment, including the political economist Adam Smith, the chemist Joseph Black (who discovered CO_2), James Watt (of steam engine fame), James Hall (who did some of the first experiments in geology), and the mathematician who would eventually become his biographer, John Playfair.

Hutton presented his paper *Concerning the Systems of the Earth* to the Geological Society in 1785, which was then published in 1788 as *Theory of the Earth*.[8] This paper also formed volume one of his two-volume *Theory of the Earth, with Proofs and Illustrations,* published in 1795.[9] Because of Hutton's obtuse prose (Steven Jay Gould, evolutionary biologist and historian, said he was a "lousy" writer),[10] this book received little attention except from opponents who supported Abraham Werner's views. Hutton's views were popularized by his friend John Playfair in *Illustrations of the Huttonian Theory of the Earth*, published in readable prose in 1802, after Hutton's death.[7]

At the beginning of his 1788 paper, Hutton viewed the globe as a machine constructed on chemical and physical principles with the purpose to support animals and humans. He recognized the solid earth, the seas, and the atmosphere as being interconnected: "it is in the manner in which these constituent bodies are adjusted to one another and the laws of action by which they are maintained in their proper qualities and respective departments that form the theory of the machine we are now to examine."[8] This statement is a remarkably modern form of what is now termed Earth System Science, a perspective that emphasizes the interdisciplinary nature of the Earth Sciences.[11] Only von Humboldt in his *Cosmos* (1856) came close to such a modern position.[12]

Hutton is best known, however, as the chief proponent of the Plutonist or Vulcanist school, whereby denudation (erosion) eventually reduces the continents to sea level and ocean sediments are rejuvenated back onto the continents by igneous activity at depth, but exactly how igneous activity caused rejuvenation was not ventured into.[9] How fossiliferous oceanic sediments were lofted onto mountain tops persisted as the single most important geological puzzle well into the late twentieth century (see Chapter 5).

Hutton recognized the importance of unconformities as reflecting deformation and folding of sediments followed by erosion and renewed deposition in a cyclic fashion (Figure 1.1). Hutton's overall view was that the Earth's history was cyclic (or repetitious) rather than historical (or progressive), and for this view he has been criticized.[10] In defense of Hutton, it should be pointed out that the geologic timescale had not yet been established at the time he was

FIGURE I.I Hutton's unconformity on Arran Island, western Scotland. Devonian Old Red Sandstone dipping moderately to the right overlies steeply dipping Dalradian (late-Precambrian) schist. Hutton visited the site in 1787.

writing, so the idea that geology would become a historical science did not yet exist. Hutton also recognized granite as an intrusive igneous rock, commonly younger than the surrounding rocks.[13] At the time, igneous rocks were called "primitive" rocks by the Neptunists and thought to be the oldest rocks of all.

The most quoted sentence from Hutton's work is the last sentence of his 1788 paper: "There is no vestige of a beginning and no prospect of an end." This was interpreted to imply there was no creation, which drew accusations of atheism from colleagues.[14] Possibly, for insurance against such attacks, Hutton peppered his text with statements such as "Devine Wisdom" and "work of infinite power and wisdom." Having spent two years in Paris, the religious views of Hutton, although not known, may have been similar to his more irreligious Enlightenment colleagues on the continent.

Abraham Werner (1749–1817). Born in Saxony, in eastern Germany, to a family with a long history in mining, Abraham Werner received his first formal education from his father who encouraged his interest in mineralogy.[16] After schooling, he took a position in his father's iron foundry. He decided to study mineralogy and mining as a career at the mining school of Freiburg, and then went on to Leipzig University, where he also studied law. In 1774, he published a paper entitled "On the External Characteristics of Fossils," a paper entirely focused on mineralogy rather than biological fossils as the word would come to be defined. In 1775, he was offered a position as a teacher at the School of Mines at Freiberg where he remained for more than forty years. Werner was methodical and orderly, but he did not take to writing. In fact, he published less as he got older – his ideas and subsequent fame were spread chiefly by word of mouth of his students.[16,17]

Werner was very popular as a teacher and eventually he drew students from across Europe to study his "geognosy," as he called geology. His famous students include Robert Jameson (1774–1854), later to become professor at Edinburgh; the polymath explorer Alexander von Humboldt (1769–1859); and Leopold von Buch (1774–1853), the Alpine geologist. Werner and his students were the chief proponents of the Neptunist school, in which all rocks were thought to be derived from a primeval global ocean through chemical precipitation (e.g., salt, gypsum, and limestone) or physical precipitation (e.g., shales, sandstones, and graywackes), including the igneous rocks (basalt and granite) as well. Eventually, even many of his own students saw that Werner's Neptunism was invalid on a global scale, and they gradually accepted the Plutonist school concept that igneous rocks formed from molten rocks (fusion), as Hutton had long maintained. Archibald Geike noted in his 1905 book that Werner made important contributions to mineralogy, but he was "disastrous to the higher interests of geology."[17] Lyell was equally unimpressed, saying in his *Principles*: "Werner's theory was original, but it was extremely erroneous." On the other hand, a glowing review of Werner's legacy can be found in the *Complete Dictionary of Scientific Biography.*[15] The Neptunist–Plutonist controversy is outlined in more detail in Chapter 4.

Georges Cuvier (1769–1832). Born in France, Cuvier was one of Europe's most influential scientists of his day in the fields of zoology, paleontology, and geology.[18] He held the chair of comparative anatomy at the French National Museum of Natural History, Paris. Historically, he was on the losing side of two important eighteenth- and nineteenth-century controversies: the transformation of one species to another (first championed by Lamarck and later by Darwin) and his catastrophist view of geologic biohistory.[18] He developed the fields of comparative anatomy and biological classification and is credited with the first report on extinctions. His ability to identify different species from fossil bones was unmatched in Europe. His studies of modern organisms and of fossils led him to conclude that many fossil species represented ancient life and were extinct. For example, he recognized, through detailed anatomical comparison, that the Indian and African elephants were different species, and that the bones of these modern elephants were different from the fossil bones of the woolly mammoth and mastodons, which were both extinct. Extinction at the time was thought to be impossible because "Almighty Wisdom" would not permit organisms that had been divinely created to die out. Human fossils had not yet been recognized in the diluvial (glacial) sediments to Cuvier's satisfaction, so in his view, extinctions were not attributable to hunting by mankind. Whether the woolly mammoth became extinct during the Pleistocene epoch due to overhunting or climate change is still debated today.[19] His recognition of extinctions led him to a catastrophist view of Earth's biohistory.

Cuvier was mainly a "cabinet" scientist who built a world-class collection of museum specimens, but he was less aware of the rapid developments taking place in geology at the time, particularly in stratigraphy and the work of William Smith (1769–1839) in England.[17] He augmented his museum collection by asking "savants and amateurs"[18] to submit fossil specimens to him, and in return he would identify them, which produced a tremendous response from the international community. His important work with Alexandre Brongniart, his colleague at the Paris museum, on the Paris basin sediments and their fossils was one of his few field-based publications.[20]

His most important work *Recherches sur les ossemens fossiles* (*Researches on Fossil Bones*) was published in four volumes in 1812. The first volume, which was written last, is known as *Discours Pré-liminaire* and was written for a general audience as a preface to the subsequent and more academic volumes, and became a very popular book translated into several European languages.[21] He recognized several different extinction events ("revolutions" in Earth history), but he thought the most recent extinction event occurred about 6,000 years ago and corresponded to the so-called diluvial deposits (now known to be glacial in origin), associated at the time with the biblical flood. There is no evidence, however, in his *Discours* that he was a biblical literalist, which in any case would have been highly unusual for a French Enlightenment scientist. He emphasized that processes operating today were insufficient to cause his "Earth revo-lutions," which was clearly a swipe at Hutton and Playfair before him, both of whom he referenced in his *Discours*. Cuvier died a few years after Lyell's *Principles* were published.

Cuvier was a gifted writer and illustrator, and a highly organized scientist. Lyell, after visiting Cuvier at his Paris museum, in a letter to his sister Marianne, marveled at Cuvier's efficient organization and work habits, noting that when Cuvier was working on a manuscript he placed all references on that topic in a single room so that he had everything at hand.[22] Lyell also notes that Cuvier's assistants "save him every mechanical labor, find references et cetera, are rarely admitted to his study, receive orders and speak not." (see Box 1.1) According to Lyell, Cuvier's library was also perfectly ordered according to zoological subject. Cuvier's intellectual heft gave sub-stantial weight to the catastrophist school of thought, especially with regard to species extinctions.

William Smith (1769–1839). William Smith was born in the village of Churchill, Oxfordshire, in southeast England. His father, who died when William was seven, was the village blacksmith.[23]

William attended the village school, where he learned to read and write, until the age of eleven. At the age of eighteen, he received a job offer as an assistant to a land surveyor, which gave him the opportunity to travel around the country. He became interested in the local strata and their fossils in different areas. He became involved in surveying for the Somerset Coal Canal in 1795 and was employed by the canal company from 1794 to 1799. This allowed Smith to examine strata over the course of the canal route. By 1796 he recognized that lithologically similar strata contained the same fossil assemblage, and he began making colored geological maps of local areas involving Triassic and Jurassic rocks. In 1804 he leased a house in London where he displayed his fossil collection arranged according to age. This collection was examined by members of the newly formed Geological Society in 1808.

In 1815 he produced his colored map of England and Wales (and parts of Scotland), which was the first geologic map of an entire country, covering an area of 65,000 square miles (104,000 square kilometers). The map was also large, measuring 2.6 m by 1.8 m. (8.5 ft by 5.9 ft). He sold about 370 copies at five guineas each (about five pounds sterling), but because of production costs (each was colored by hand), he made little money. A business venture involving the quarrying of building stone was a failure, and he ran up large debts. He sold his fossil collection to the British Museum for a modest sum. He continued to travel wherever his land-surveying jobs and civil-engineering projects took him. Adam Sedgwick (see in this chapter), president of the Geological Society, awarded Smith the first Wollaston medal in 1831, the Society's highest award. The government provided him a modest annuity of one hundred pounds sterling in recognition of his services to the country for producing his 1815 map, which had important economic and practical implications. Smith died at the age of seventy while traveling. The 200th anniversary of his map was celebrated in 2015 by an article in *Nature* magazine.[24]

William Buckland (1784–1856). Born in Devonshire, England, William Buckland appears to have been interested in natural history from an early age.[25] He studied theology at Christ's College, Oxford, and was ordained in 1809. At Oxford, he met others interested in geology and fossils, including William Conybeare (see in this chapter). From 1808 to 1815, he made geological excursions in England and also to Europe. He was elected reader in mineralogy at Oxford in 1813, and also became a fellow of the Geological Society of London. He was president of the Geological Society 1824–1825, and then again 1840–1841.

A treatise on geology and mineralogy by Buckland, sponsored by a wealthy patron (Earl Bridgewater), published in 1837, was entitled *The Bridgewater Treatises on the Power, Wisdom and Goodness of God as Manifested in the Creation.*[26] The final section of volume one was entitled, "Geological Proof of a Deity," where Buckland argued that the order seen in geology is proof of the existence of one "Supreme Creator."[26] The treatise nevertheless represents a major contribution to geology and paleontology, and was accompanied by many excellent illustrations (Figure 1.2). Although he held creationist views, Buckland may not have been a biblical literalist, at least not in his later years. In his presidential address of 1841 (at age fifty-seven), he accepted the evidence in Scotland and the Alps, presented to him in the field by Louis Agassiz (see in this chapter), that so called diluvial sediments were deposited by ice rather than the biblical flood.[27] Buckland was forty-six years old when *Principles* was first published (Table 1.1).

Adam Sedgwick (1785–1873). Born in Yorkshire as son of a clergyman, Sedgwick studied theology and mathematics at Cambridge.[28] He stayed on at Cambridge as a tutor after graduation; he was ordained in 1817, and he became professor of geology at Cambridge in 1818 at the age of thirty-three. He began his geological research in the Lake District of northern England in 1822, where he met and befriended the poet William Wordsworth. Conybeare and Phillips's *Geology of England and Wales* was published in the same year, and Sedgwick likely had a copy of this book. His first journey with Roderick Murchinson (see in this chapter) was to Scotland in

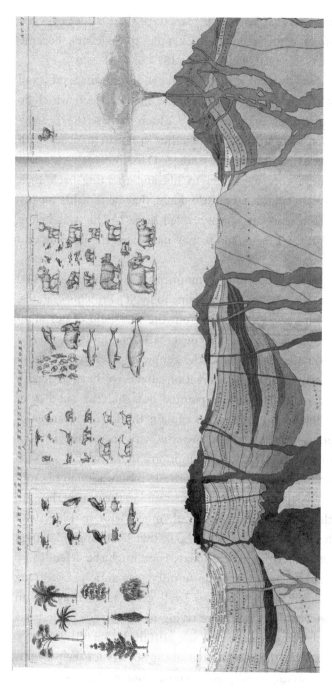

FIGURE 1 2 A small part of Plate I from William Buckland's *Geology and Mineralogy Considered with Reference to Natural Theology* (1837). The geologic cross section is an idealized section through the Earth's crust showing granite at depth overlain by primary, secondary, and tertiary strata together with extinct volcanoes and active volcanoes. Animals and plants based on the fossil record are shown above.

Source: Buckland, W. 1837. *The Bridgewater Treatises on the Power, Wisdom and Goodness of God as Manifested in Creation. Treatise VI. Geology and Mineralogy Considered with Reference to Natural Theology* (2 vols.). Pickering, London.

1826, a collaboration that lasted ten years until their fallout over the Silurian-Cambrian controversy (see Chapter 2). Sedgwick became president of the Geological Society from 1829 to 1831.

Sedgwick was a strong critic of Lyell's uniformitarianism,[27] and he was also hostile to Darwin's *Origin of the Species* (1859), which he labeled as "unflinching materialism."[29] Much earlier (1831), Sedgwick and Darwin were in the field together in northern Wales with Sedgwick acting as mentor to the younger Darwin. Sedgwick asked Darwin to trace a course parallel to his and to bring back rock samples. Darwin noted, "I have little doubt that he did this for my good, as I was too ignorant to have aided him."[30] This was before Darwin sailed on the Beagle, and it may have been his introduction to geology.

Speaking on the doctrine of uniformity in his second (1831) address to the Society, Sedgwick, referring to Lyell's *Principles*, states: "We must banish *apriori* reasoning from the threshold of our argument and the language of theory can never fall from our lips with any grace or fitness, unless it appear as the simple enumeration of those general facts, with which by observation alone, we have at length become acquainted."[28] He further pointed out that the subtitle of Lyell's book ("[b]eing an attempt to explain the former changes of the Earth's surface by reference to causes now in operation") is itself an *a priori* assumption. (*A priori* assumptions are those that are self-evident and require no substantiation). Lyell's assumption in his subtitle was that geologic processes were the same in the past as they are today. It is clear in a letter from Lyell, much later in 1837, to a reviewer of the second edition of his book that Lyell misunderstood the meaning of the term *a priori*. Today, Lyell might be defended by saying his initial assumption of uniformity was simply a working hypothesis. Sedgwick's invective to Lyell gives some insight into how these pioneers of geology were trying to build a scientific basis for a nascent science, a science that was still in the "cradle."

In summarizing the publications of the Society for the previous year in his 1831 address, Sedgwick pointed out geologic situations

where large dislocations have juxtaposed strongly contorted formations against horizontal strata, concluding that these situations indicated not slow events, but "violent events of short duration."[28] The
Reverend Sedgwick, although a catastrophist, was no scriptural geologist, and in his first annual address to the Society (1830), he reviewed a
new book by the Irishman Andrew Ure, A New System of Geology,
which attempted to reconcile geology with the biblical account of
Earth's origins.[31] His review of the book was very harsh. He pointed
out gross and elementary errors of fact and even plagiarism, leaving
no doubt he had no time for such "monuments of folly." From his
research in Wales in the 1830s, Sedgwick is responsible for our current
Cambrian system (see Chapter 2).

William Conybeare (1787–1857). Born in London, he was the
son of the Reverend William Conybeare and was educated at West-
minster school at Oxford.[32] He married in 1814 at age twenty-seven,
served as vicar in Devonshire in 1836, and became a dean in 1845. He
was an early member of the Geological Society of London (1811), and
was elected a fellow of the Geological Society in 1832. He was a close
associate of William Buckland with only three years separating them
in age; both were schooled at Oxford and both were clergymen.

His most important geological work is Outlines of the Geology
of England and Wales (Conybeare and Phillips, 1822), which is a
detailed description of geologic formations down to the Carboniferous
and includes a geological map of England and Wales with selected
cross-sections of highly folded regions (Figure 1.3). Phillips was the
publisher and most of the geologic work is attributed to Conybeare.[33]
Unconventionally, the book begins with descriptions of the youngest
strata first – alluvial and underlying diluvial deposits of sand and
gravel, the latter widely interpreted to represent the universal Noa-
chian Deluge, or biblical flood (see letters from Lyell in Box 1.1). Part
of the introduction to Outlines tried to reconcile the biblical account
of the flood with the stratigraphic record; it entertained that the
length of days of creation in the Bible may have been metaphorical,
rather than lasting twenty-four literal hours. Conybeare asked whether

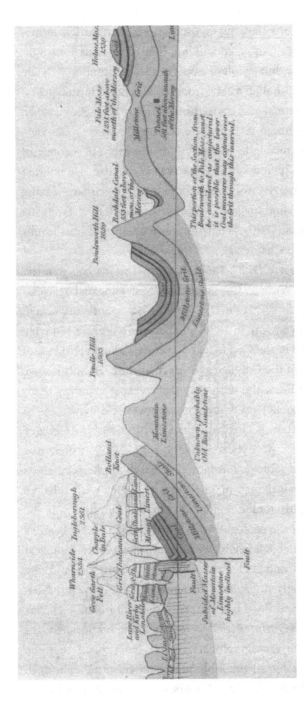

FIGURE 1.3 Part of a cross-section from the Irish Sea in Cumberland to the North Sea in Durham showing folded Carboniferous strata from Figure 3 of Conybeare and Phillips, *Outlines of the Geology of England and Wales*, 1822. Mountain limestone occurs in the core of the anticline and the coal measures outcrop in the core of the synclines. Millstone grit and limestone and shale underlie the coal measures. The Old Red Sandstone (Devonian) underlies the mountain limestone and was thought to be part of the Carboniferous system. Original horizontal scale: 10 miles to an inch; original vertical scale: 400 feet to 1/8 inch, corresponding to a vertical exaggeration of 16.5.

Source: Conybeare, W. and Phillips, W. 1822. *Outlines of the Geology of England and Wales.* Phillips, London.

there enough time to explain the great thickness of pre-diluvial strata (called Secondary strata) in the period between the Creation and the biblical deluge? Conybeare was not a true biblical literalist, he was more flexible— but he was clearly trying to reconcile scripture with the wealth of new stratigraphic information at hand. Also, in the introduction to *Outlines*, Conybeare attempted to test Werner's Neptunist hypothesis, which predicts formations on either side of a sedimentary basin should occur at the same elevation. He finds that the Neptunist idea was plausible for shallow-dipping strata in southeast England, but not for the Alpine region. Based on formations in the Alps, he also argued against Lyell's uniformity principle.

Conybeare was a catastrophist and in an open letter to Lyell in *Philosophical Magazine* (1830), he challenged Lyell's assumption of uniformity in intensity of nature's laws.[5] In a footnote in the same letter, he accused Lyell of plagiarism of his own words, to which Lyell replied to indicate that they were not on friendly terms as the pleasantries in their publications would suggest (see also Box 1.1).[34] (Lyell's defense was that the quotes he used were from classical literature and were so hackneyed that Conybeare deserved no acknowledgment for them.) For good measure, Conybeare further commented on Hutton's final sentence in *Theory of the Earth* (quoted earlier): "I continue to regard it as one of the most gratuitous and unsupported assertions ever hazarded." He made clear, however, that this objection was not on moral grounds, but on scientific grounds. We will return to Conybeare's objections to Lyell's uniformity principle later in the chapter.

Roderick Murchison (1792–1871). Born in Tarradale in the Scottish Highlands to a landowning family, Murchinson's father died when he was four.[36] He was educated at a military school and at age sixteen saw active service in 1808 in the Spanish Penninsular war. In 1815 he married and soon afterward resigned his military position. He traveled in Italy (1816–1818), and on his return he sold his estate (making him independently wealthy), and he took up fox hunting in the Highlands.

In 1824 he settled in London, and with encouragement from his wife and William Buckland he became interested in geology. He was elected to the Geological Society of London in 1825, and he focused on stratigraphy using Conybeare and Phillips's 1822 book as his template. In 1828 he traveled with Lyell in France and northern Italy, and also with Sedgwick, where they undertook studies of the eastern Alps; they published their results in a paper in 1832.[37] In 1831 Murchison commenced his study of lower Paleozoic sediments in Wales, which later became his famous Silurian System (1835). This work in Wales was also the source of the split between him and Sedgwick, which lasted the rest of their lives (see Chapter 2).

Murchison was indefatigable; his geologic career is marked by long arduous trips in Europe and Russia over a period of twenty years, seeking out new stratigraphic sections – a testament to his physical stamina and endurance (he lived to be seventy-nine). After one of his trips to Russia, he reported to the *Philosophical Magazine* in 1841 that he covered between thirteen and fourteen thousand miles. Together with Sedgwick, he is responsible for the names of three of the eleven Phanerozoic geologic periods (the Permian, Silurian, and Devonian). In 1855 he succeeded De la Beche (1796–1855) as director of the Geologic Survey of Britain, which essentially gave him the upper hand in the Silurian-Cambrian dispute with Sedgwick – survey maps omitted or reduced areas underlain by Sedgwick's Cambrian strata. He opposed Darwin's evolution theory and also Agassiz's glacial theory, and became increasingly intolerant of innovation in geology as he aged.[34] Martin Rudwick has pointed out that when involved in scientific disputes, he invariably used military metaphors in the dispute, possibly reflecting his military experience.[3] On account of his independent wealth, Murchison was a major figure in London's high society.

Charles Lyell (1797–1875). Born in Scotland to a well-off family, Lyell attended school in Sussex, southern England, to where his family had moved.[38] At age twenty he enrolled in Exeter College, Cambridge, and in 1817 he attended lectures on mineralogy and geology given by the Reverend William Buckland. Lyell toured Europe in 1818 in a carriage with his parents and sisters where they crossed the Alps. Whenever he traveled, which was often, he took an interest in the local geology. He visited Paris in 1823, where he studied French (the language of the sciences at the time), and he attended free lectures on science, meeting at the same time many of the famous naturalists of the day such as Cuvier, Brongniart, von Humboldt, and Prévost. He also learned German and Italian so that he could read the geologic literature widely.[38] He already knew Latin and Greek.[2]

By 1829, in letters to Murchison, Lyell was beginning to formulate an idea that would occupy him all his life, namely that present-day natural forces could explain all geological events and there was no need for catastrophes in the geologic past (see Box 1.1). Cuvier and Brongniart had published an important study of the sediments and fossils of the Paris basin in 1811.[20] They saw that freshwater shells alternated with marine shells, and they interpreted this to be due to revolutions in Earth history, but Lyell thought the pattern could also be explained by slow alternation between freshwater and marine conditions. He saw a similarity between present-day Scottish lake sediments that he had studied earlier to the older Paris sediments, and saw no need to invoke Earth revolutions. His travels to volcanic areas in Italy led him to the same conclusion.[1]

Lyell practiced law from 1825 to 1827 but poor eyesight suggested he needed a different career, and he chose geology. He was elected a fellow of the Geological Society in 1826. In 1828 he visited the Auvergne volcanic region in south-central France with Murchison. In 1831 he was appointed professor at King's College London where he lectured in geology, but did not remain there for long. His interests turned to publishing a book on geology, possibly for better financial gain compared to that of academia. He published the first edition of his most important work, *Principles of Geology*, in three volumes between 1830 and 1833; it was very popular among the general public and underwent twelve editions (the last edition being posthumous).

For Lyell, not only were presently operative causes able to explain past events, but he implausibly maintained their *intensity* was the same as today. This made his book quite controversial, especially among catastrophists such as Sedgwick and Conybeare. Lyell also did not accept that the Earth was undergoing long-term cooling, which would have invalidated his uniformity of nature assumption. Lord Kelvin, the English physicist (1824–1907) repeatedly pointed out that Lyell's view violated the second law of thermodynamics (see Chapter 3).

Charles Darwin and Lyell became good friends: "I saw more of Lyell than of any other man both before and after my marriage. His mind was characterized, as it appeared to me, by clearness, caution, sound judgment and a good deal of originality."[40] Lyell welcomed Darwin's new theory on coral reefs but was much less enthusiastic about Darwin's natural selection theory in *Origin of the Species* (1859). For Lyell, the evolution of the species implied a progressive path which was inconsistent with his repetitive cyclic approach and his concept of uniformity in nature.

Louis Agassiz (1807–1873). He was born in Switzerland and died in Cambridge, Massachusetts.[41] His early contribution was in ichthyology (study of fossil fish), but Agassiz later gained widespread fame for development of the concept of an Ice Age: namely, that the northern hemisphere had been covered in thick ice during the Pleistocene epoch. He attended the universities of Zurich, Heidelburg, and Munich. Agassiz's degrees include a doctorate in science and a degree in medicine conferred in 1830 from Munich. His monograph on fossil fish gained the attention of Cuvier. He accepted a professorship at Neuchâtel in 1832, and under mentorship from Johann von Charpentier (a student of Werner) studied glacial deposits of the Swiss Alps. Agassiz persuaded William Buckland of his glacial theory in the Alps and the British Isles in 1838, who in turn, persuaded Lyell that extensive glaciations had occurred in these regions. After the death of his wife in 1847, he moved to Harvard University, where, after extensive travel, he extended his glacial theory to North America as well.

He was a catastrophist, believing that the Ice Age had caused extinction of species such as the woolly mammoth (see Cuvier in this chapter). Agassiz was also a strong opponent of Darwin's ideas on evolution and of Lyell's uniformity of geologic processes. However, according to marginalia in Agassiz's personal copy of the *Principles* (discovered by Steven Gould in Harvard's library), he greatly admired Lyell's book.[10]

Charles Darwin (1809–1882). Arguably the most important scientist since Newton, Darwin's theory of evolution by natural selection (*Origin of the Species*, 1859) demoted humans from "Almighty Wisdom's" finest creation to mere descendents of the lower primates. He was at first a geologist and only later became a biologist. His academic career was inauspicious from the start and gave no inkling of his genius, which turned out to be in three main areas: geology, evolution of species, and botany.[42] His most important *geological* contribution is on the origin of coral reefs.[42]

Darwin's mother died young, and his sisters took care of him in his early years.[42] He was dubbed "a slow leaner" in school. His father sent him to Edinburgh University to study medicine, but observing operations without anesthesia convinced him medicine was not for him. He was then sent to Cambridge University to be a clergyman, and graduated in 1831 with a poor degree. Commenting on his academic experience, he noted later: "wasted as far as the academical studies were concerned as Edinburgh and school."[42] At Cambridge he did, however, meet Adam Sedgwick and other friends who gave him an interest in geology.

The most important event of his life turned out to be the invitation, after graduation from Cambridge, to accompany captain Robert Fitzroy on the voyage of the *Beagle* (1831–1836) as an unpaid naturalist to explore coastal South America, including some Pacific coral islands. The origin of coals reefs was one of the specific goals of the expedition. Lyell had a chapter in his recently published *Principles* (1830–1833) on coral reefs, which focused on atolls with their circular shape and shallow lagoons. (Darwin had a copy of Lyell's book aboard the *Beagle*.) Lyell surmised that atolls were formed on the rims of submerged volcanoes, their circular shape and size mimicking the volcanic crater.

On return from his five-year voyage Darwin's first major publication was *The Structure and Distribution of Coral Reefs* published in 1842.[43] Darwin's theory was global in scale and far exceeded in explanatory ability Lyell's earlier theory, showing how coral reefs evolved in time from fringing reefs, to barrier reefs, and finally to atolls, due to gradual volcanic subsidence (Figure 1.4). This theory conformed to Lyell's uniformitarianism, and Lyell welcomed it.

DISCUSSION

As a generalization, catastrophism was thought to be historical in its approach to the geologic and fossil record, whereas the uniformity position was ahistorical, being thought of as cyclic rather than progressive with time.[10] This cyclic nature of geologic processes was first emphasized by Hutton and later by Lyell; the uniformity position was also repetitious, whereas catastrophism did not see repetition in the geologic record.[44]

In a private letter to Lyell on the eve of his departure to visit America in 1841, Conybeare summarized the main points against Lyell's uniformity in nature position.[6] No reply from Lyell has been found – he may have been pre-occupied with preparation for his upcoming trip. The letter was prompted by Lyell having sent Conybeare the sixth edition of *Principles*. Conybeare essentially summarized the arguments he had already published in *Philosophical Magazine* in several installments in the 1830s.[5]. He agreed with the position of the uniformity of the laws of nature and general physical causes, but he argued that "different conditions at different times materially modified their intensity."[5] This was the major difference between catastrophists and uniformitarians at the time.

Citing the fossil record seen in younger to older sediments – namely, man (recent strata), mammalia (tertiary strata), reptiles (secondary strata), and fish (primary strata) – Conybeare concluded

(a)

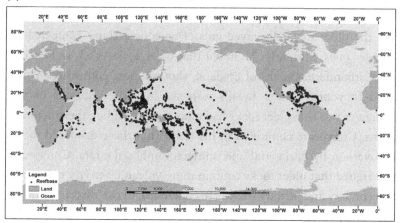

(b)

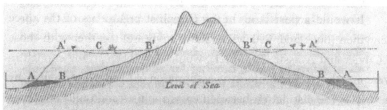

[No. 4]

A A—Outer edge of the reef at the level of the sea.

B B—Shores of the island.

A′ A′—Outer edge of the reef, after its upward growth during a period of subsidence.

C C—The lagoon-channel between the reef and the shores of the now encircled land.

B′ B′—The shores of the encircled island.

N.B. In this, and the following wood-cut, the subsidence of the land could only be represented by an apparent rise in the level of the sea.

FIGURE 1.4 (a) Darwin provided a global map of coral reefs similar to this map in his 1842 publication,[43] color-coded according to reef type: fringing reefs, barrier reefs, and atolls. Reefs along continental margins are mainly fringing reefs and barrier reefs, whereas isolated ocean reefs are mainly atolls with lagoons.

Source: NOAA.

(b) Fringing reefs are the first to form on the shores of a volcanic island (A–B). As the island subsides, the reef grows upward to become a barrier reef (A′–B′) with a lagoon (C). Eventually the island subsides below sea level, and an atoll develops on the volcanic crater rim.

Source: Ref. 43: Darwin, 1842.

"an arrangement of progressive organization" was apparent in the history of life.[5] Referring to igneous and metamorphic rocks, he noted that the older rocks displayed greater igneous activity compared to younger ones. Similarly, he noted that older rocks, such as Silurian and Carboniferous rocks of England, showed more contortions compared to younger ones. Lyell's response in earlier editions of the *Principles* was that older rocks had more time to endure these deformations. Conybeare countered in his letter that these deformed rocks are overlain by horizontal and unmetamorphosed strata. Conybeare also argued that older rocks contain more volcanic activity compared to secondary and tertiary rocks. He concluded, "I can discern nothing like this regularly recurring series of uniform events."[5] Lyell's response in his *Principles* was that the fossil record was incomplete and that the stratigraphic record was also very incomplete.

It would appear that the catastrophist arguments of the three clergymen (Buckland, Sedgwick, and Conybeare) together with those of Cuvier and Agassiz (Table 1.1), would have been strong enough to overcome the *a priori* uniformity assumption of Lyell. Nevertheless, by the 1850s, the distinction between catastrophists and uniformitarians began to blur.[45] As Gould pointed out in his 1965 paper, the uniformitarian doctrine is no longer relevant to modern science and this was true by the end of the nineteenth century.[46] In the next chapter, the construction of the geological timescale is addressed, and we will meet some of the same players again in a different context.

REFERENCES

1. Holmes, A. 1913. *The Age of the Earth*. Harper & Brothers, London.
2. Lyell, C. 1830–1833 (1997). *Principles of Geology* (3 vols.). Penguin, London.
3. Whewell, W. 1832. Review of principles of geology (v. 2). *Quart. Rev.*, v. 47, 103–132.
4. Rudwick, M. 2008. *Worlds before Adam*. Chicago University Press, Chicago.
5. Conybeare, W. D. 1830. An examination of those phenomena of geology, which seem to bear most directly on theoretical speculations. *Philosophical Magazine*, Ser. 2, v. 8, 401–406.

6. Rudwick, M. 1967. A critique of uniformitarian geology: a letter from W. D. Conybeare to Charles Lyell, 1841. *Proceedings American Philosophical Society*, v. 111, 272–287.

7. Playfair, J. 1802 (1956). *Illustrations of the Huttonian Theory of the Earth.* University of Illinois, Urbana.

8. Hutton, J. 1788. Theory of the Earth: or an investigation of the laws observable in the composition, dissolution, and restoration of land upon the globe. *Transactions of Royal Society Edinburgh*, v. 1, 209–304.

9. Eyles, V. A. 2008. James Hutton. *Complete Dictionary of Scientific Biography*, v. 6, 577–589. Charles Scribner's Sons, Detroit.

10. Gould, S. J. 1987. *Time's Arrow, Time's Cycle.* Harvard University Press, Cambridge, MA.

11. Skinner, B. J. and Porter, S. C. 1995. *The Blue Planet: An Introduction to Earth System Science.* Wiley, Hoboken, NJ.

12. Jackson, S. T. 2009. Alexander von Humboldt and the general physics of the Earth. *Science*, v. 324, 596–597.

13. Hutton, J. 1794. Observations on granite. *Transactions of Royal Society Edinburgh*, v. 3, 77–81.

14. Whewell, W. 1831. Review of *Principles of Geology* (v. 1). *British Critic, Quarterly Theological Review, and Ecclesiastical Record*, v. 9, 180–206.

15. Ospovat, A. 2008. Abraham Werner. *Complete Dictionary of Scientific Biography*, v. 14, 256–264. Charles Scribner's Sons, Detroit.

16. Geike, A. 1905. *The Founders of Geology.* Macmillan, London.

17. Von Zittel, K. A. 1901. *History of Geology and Paleontology.* W. Scott, London.

18. Rudwick, M. 2008. Georges Cuvier. *Complete Dictionary of Scientific Biography*, v. 20, 221–227. Charles Scribner's Sons, Detroit.

19. Guthrie, R. D. 2006. New carbon dates link climatic change with human colonization and Pleistocene extinctions. *Nature*, v. 441, 207–209.

20. Cuvier, G. and Brongniart, A. 1811. Essai sur la géographie mineralogique des environs de Paris, avec une carte géognostique. *Mémoires de la classe des sciences mathematiques et physiques et de l'institut Imperial de France*, 1–278.

21. Cuvier, G. 1831. *A Discourse on the Revolutions of the Surface of the Globe and the Changes Thereby Produced in the Animal Kingdom.* Carey and Lea, Philadelphia.

22. Lyell, K. M. (ed.). 1881. *Life, Letters and Journals of Charles Lyell.* J. Murray, London.

23. Eyles, J. M. 2008. William Smith. *Complete Dictionary of Scientific Biography*, v. 12, 486–492. Charles Scribner's Sons, Detroit.

24. Sharpe, T. 2015. The birth of the geological map. *Nature*, v. 347, 230–232.

25. Cannon, W. F. 2008. William Buckland. *Complete Dictionary of Scientific Biography*, v. 2, 566–572. Charles Scribner's Sons, Detroit.

26. Buckland, W. 1837. *The Bridgewater Treatises on the Power, Wisdom and Goodness of God as Manifested in Creation. Treatise VI. Geology and Mineralogy Considered with Reference to Natural Theology* (2 vols.). Pickering, London.

27. Buckland, W. 1841. Annual address. *Proceedings of Geological Society of London*, v. 3, part 2, 327–337.

28. Rudwick, M. 2008. Adam Sedgwick. *Complete Dictionary of Scientific Biography*, v. 12, 275–279. Charles Scribner's Sons, Detroit.

29. Sedgwick, A. 1831. Annual address. *Proceedings of Geological Society of London*, v. 1, 271–317.

30. Sedgwick, A. 1860. Objections to Mr. Darwin's theory of the origin of species (written on behalf of the archbishop of Dublin). *Spectator*, March 24, 285–286.

31. Craig, G. Y. and Jones, E. J. (compilers). 1982. Sedgwick and Darwin in the Field. *A Geological Miscellany*, p. 134. Princeton University Press, New Jersey.

32. Sedgwick, A. 1830. Annual address. *Proceedings Geological Society London*, v. 1, 187–212.

33. Rudwick, M. 2008. William Conybeare. *Complete Dictionary of Scientific Biography*, v. 3, 395–396. Charles Scribner's Sons, Detroit.

34. Conybeare, W. and Phillips, W. 1822. *Outlines of the Geology of England and Wales*. Phillips, London.

35. Lyell, C. 1831. Reply to note in the Rev. Conybeare's paper entitled, "An examination of those phenomena of geology which seem to bear most directly on theoretical speculations." *Philosophical Magazine*, series 2, v. 9, 1–3.

36. Rudwick, M. 2008. Roderick Murchison. *Complete Dictionary of Scientific Biography*, v. 9, 582–585. Charles Scribner's Sons, Detroit.

37. Sedgwick, A. and Murchison, R. 1832. A sketch of the structure of the eastern Alps. *Transactions of Geological Society of London*, series 2, v. 3, 301–420.

38. Rudwick, M. 2008. Charles Lyell. *Complete Dictionary of Scientific Biography*, v. 8, 563–576. Charles Scribner's Sons, Detroit.

39. Rudwick, M. 1998. Lyell and the *Principles of Geology*. In *Lyell: the past is the key to the present* (Blundell, D. J. and Scott, A. C., eds.). Geological Society London special paper 143. London.

40. Craig, G. Y. and Jones, E. J. (compilers). 1982. Darwin on Lyell. *A Geological Miscellany*, 135. Princeton University Press, New Jersey.

41. Lurie, E. 2008. Louis Agassiz. *Complete Dictionary of Scientific Biography*, v. 1, 72–74. Charles Scribner's Sons, Detroit.

42. Bowler, P. 2008. Charles Darwin. *Complete Dictionary of Scientific Biography*, v. 20, 242–249. Charles Scribner's Sons, Detroit.

43. Darwin, C. 1842. *The Structure and Distribution of Coral Reefs*. Stewart and Murray, London.

44. Hooykaas, R. 1975. Catastrophism in geology, its scientific character in relation to actualism and uniformitarianism in *Philosophy of Geohistory* (Albritton, C. C. ed.). *Benchmark Papers in Geology*, v. 13, 310–356.

45. Wilson, L. G. 1980. Geology on the eve of Charles Lyell's first visit to America, 1841. *Proceedings of American Philosophical Society*, v. 124, 168–202.

46. Gould, S. J. 1965. Is uniformitarianism necessary? *American Journal of Science*, v. 263, 223–228.

2   Toward a Geologic Time Scale

I have previously pointed out that the age of the radioactive minerals can be calculated from the amount of helium contained in them.

— E. Rutherford, 1906[19]

INTRODUCTION

The geologic time scale is the fundamental tool geologists use to unravel Earth's history. It allows investigation into the development of life over time, including mass extinctions, repeated continental collisions and mountain-building events, as well as past environmental changes. It is a roadmap – without which geologists would be lost in what John McPhee referred to as "deep time."[1] When nineteenth-century geologists were struggling with the extent of geologic time based on the known sedimentary record, they calculated that 100 million years of sedimentation may have been required (see Chapter 3). Early in the twentieth century, after the discovery of radioactivity, the first uranium-lead ages indicated that the Cambrian could be as much as 500 million years old, and geologists now had an "embarrassment of time" on their hands. Actually, the Cambrian represents only about 12 percent of geologic time, with the Precambrian corresponding to 88 percent of geologic time.

The British geologist Arthur Holmes (1890–1965) recognized that the geologic timescale is always a work in progress as new discoveries are made and better dating techniques are developed. All geologic timescales therefore come with a date and a shelf life. The 1982 geologic timescale of Harland and others, for example, is a book of reasonable length at 130 pages.[2] The 2004 time scale of Gradstein and others is 589 pages long,[3] and the more recent 2012 version is in two volumes for a total of 1,176 pages, and this expansion is likely to continue into the future.[4] But as shown later in the chapter, over the past several decades there has been a gradual convergence toward a consensus: the main advances being refinement of the existing time-scale rather than whole-scale revision.

Construction of a modern geologic time scale involves three steps: First, a relative time scale based on stratigraphy, called the chronostratigraphic scale, and often based, at least in part, on biostratigraphy or the fossil content of the rocks. Second, a linear measure of time or elapsed duration is needed, based either on radiometric data (so-called isotopic ages), or chemical changes in ice cores with depth or reversals in the Earth's magnetic field – this is called the chronometric scale. The third step is to combine the chronostratigraphic and chronometric scales to produce a calibrated geologic timescale.[2]

The first geologic columns were not constructed in such an organized fashion as outlined above: they were developed by an ad hoc process involving unraveling the relative stratigraphic age of rock formations by individuals working independently – clergymen and professors, and what Martin Rudwick calls "gentlemen specialists." An exception to this picture is William Smith (1769–1839), a surveyor with only a village school education who produced the first geologic map of England and Wales. Some of these pioneers in geology were discussed in the previous chapter.

Today the chronostratigraphic scale is agreed upon by convention, and the International Commission on Stratigraphy plays an important role in this endeavor; the goal is to establish a global correlation of the rock record based on an agreed-upon type locality, the emphasis being on the boundaries between stratigraphic units. For example, if the Cretaceous-Tertiary boundary was not established at different locations globally, the meteorite-impact hypothesis of the dinosaur extinction at that boundary could not have been scientifically tested. In contrast, the chronometric scale is to be discovered by measurement (for example, isotopic ages) rather than by convention. Because of the poor preservation of Precambrian stratigraphy, the three steps referred to earlier apply only to the Phanerozoic Eon. In the Precambrian (comprised of the Proterozoic and Archean eons),

Table 2.1 *The structure of the modern geologic timescale*[3]

Eon	Era	Basal Age (Millions of Years)
Phanerozoic	Cenozoic	65
	Mesozoic	
	Paleozoic	542
Proterozoic	Neoproterozoic	
	Mesoproterozoic	
	Paleoproterozoic	2,500
Archean	Neoarchean	
	Mesoarchean	
	Paleoarchean	
	Eoarchean	Base not defined
		~4,600

ages are assigned to the base of eons, eras, and periods without reference to specific rock bodies (Table 2.1).

This chapter covers the historical development of the Phanerozoic timescale, which took place largely in the first half of the nineteenth century. After the discovery of radioactivity, early attempts in the twentieth century to calibrate this geologic column are summarized, and the chapter ends with a discussion of the Cretaceous-Tertiary mass extinction event 65 million years ago.

THE PHANEROZOIC

Figure 2.1 shows the progression over time of concepts of the geologic stratigraphic column up until the mid-twentieth century. The tripartite division of rock strata into Primitive, Secondary, and Tertiary represents the first rudimentary stratigraphy, and is attributed to the Italian geologist Giovanni Arduino (1713–1795), who was an inspector of mines in Tuscany and later became professor of mineralogy at Venice.[5] Primitive rocks are the oldest (unfossiliferous crystalline rocks, both igneous and metamorphic, that make up the backbone of mountain belts), followed by Secondary strata (fossiliferous limestones,

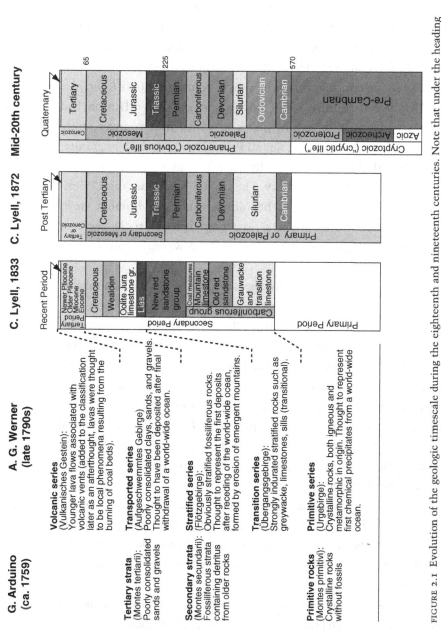

FIGURE 2.1 Evolution of the geologic timescale during the eighteenth and nineteenth centuries. Note that under the heading C. Lyell (1833), Old Red Sandstone was included in the Carboniferous before the Devonian was identified. Note also that under C. Lyell (1872), the Ordovician is absent between the Cambrian and Silurian. The Neptunist system of Abraham Werner is shown in the second column from left.

Source: Gradstein, F., Ogg, J. and Smith, W. 2004. A Geologic Time Scale. Cambridge University Press, Cambridge.

sandstones, and shales), and lastly Tertiary strata (unconsolidated sands and gravels containing plants and animal remains). German geologist Johann Lehmann (1719–1767) also recognized a similar threefold division in the Hartz Mountains of Germany.

Lyell makes a careful distinction in volume III of his *Principles* (1833) between Primitive and Primary rocks.[6] After Hutton had discovered granitic injections (dikes or veins) into younger strata (Chapter 1), it was clear that not all Primitive rocks, such as granite, were the oldest, and the term Primary was used instead – hence Primary, Secondary and Tertiary. The two terms, Primary and Primitive, are easily confused, but they differentiate two different viewpoints: that of the Neptunists, to whom igneous rocks were of sedimentary origin, and the Plutonists, for whom igneous rocks had a fusion origin (Figure 2.1).

Nicolas Steno (1638–1686) appears to be one of the first savants to recognize what modern textbooks somewhat grandiosely refer to as "the law of superposition": namely, that in a sequence of horizontal geologic strata, the overlying layers are younger than the lower layers.[7] He, furthermore, recognized marine fossils represented organic remains of existing or extinct animals and plants, and identified that they originated in the ocean and were not "sports of nature," as was the then current opinion. Born in Copenhagen, Denmark, Steno studied medicine and traveled widely in Europe before settling in Florence. He published his treatise on the geology of Tuscany in 1699, which laid the foundations of modern stratigraphy.[7]

One of the earliest geologic cross-sections was produced by the Englishman John Strachey. He recognized the stratigraphy in southwest England between the Coal Measures (Carboniferous) and the Chalk (Cretaceous), and identified Triassic red marls that unconformably overly tilted coal beds (Figure 2.2).[8]

We met Abraham Werner (1749–1817) in the previous chapter as the founder of the Neptunist doctrine, who built on these early rough stratigraphies with modifications (see Figure 2.1). The Neptunists believed that all rocks were precipitated from a universal primordial

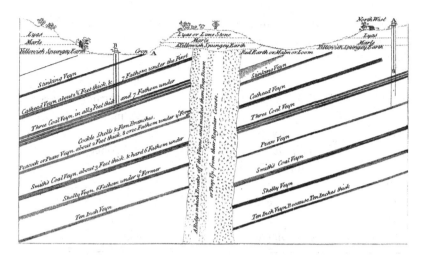

FIGURE 2.2 A cross section from southeast to northwest in coal mines in the Mendip Hills of Somersetshire, southwest England, published in 1717 by John Strachey. The coal beds (or "veins") dip to the southeast and are unconformably overlain by horizontal beds of Mesozoic age ("Lyas or limestone, Marle and Yellowish Spungey Earth"). A wide fault zone is portrayed beneath the central topographic ridge, which offsets the coal beds with down throw to the northwest.

Source: Strachey, J. 1717. A curious description of the strata observed in the coal-mines of Mendip, Somersetshire. *Philosophical Transactions Royal Society London*, v. 30, 968–974.

ocean, including granites and volcanic sills. Most of what we know about Werner's beliefs is based on his student's work (see Chapter 4), as he published little on his own.[5,6]

Table 2.2 shows the dates of recognition for the various geologic periods in the Phanerozoic, indicating that most of this intensive geologic work took place during the first half of the nineteenth century. The Phanerozoic is comprised of the Paleozoic, Mesozoic, and Cenozoic eras (see Table 2.1), corresponding to ancient life, middle life, and recent life – terms coined by the British paleontologist J. Phillips in 1841.[9] The younger fossiliferous and less-deformed rocks were deciphered first. The Carboniferous Period was recognized early on – no doubt on account of its economic importance in fueling the Industrial Revolution in England.

Table 2.2 *Recognition of phanerozoic periods by date and author*[34]

Date	Period	Author
1799	Jurassic	von Humboldt
1822	Carboniferous	Conybeare/Phillips
1822	Cretaceous	d'Omalius d'Halloy
1833	Tertiary (Neogene)	Lyell
1834	Triassic	von Alberti
1835	Silurian/Cambrian	Murchinson/Sedgwick
1839	Devonian	Murchinson/Sedgwick
1839	Pleistocene	Lyell
1841	Permian	Murchinson
1879	Ordovician	Lapworth

The geologic column in Figure 2.1 under Lyell 1872 is notable in that the Ordovician Period is absent between the Silurian and the Cambrian. The Ordovician Period was proposed by Charles Lapworth (1842–1920) in 1879 in response to the longstanding heated debate between Roderick Murchinson (1792–1871) and Adam Sedgwick (1785–1873).[10] On William Smith's geologic cross section through Wales and England of 1815, the formations of the fossiliferous Mesozoic and Upper Paleozoic formations are clearly distinguished, whereas the Lower Paleozoic of Wales is shown in purple as an undifferentiated mass of "Killas and Slate" (metamorphosed sediments). In 1831, both Murchinson and Sedgwick decided to try to unravel the stratigraphy of these largely unknown strata in Wales. Murchinson began in southern Wales and worked northward, passing through Old Red Sandstone strata knowing that the rocks he was about to study were older because they underlay the Old Red Sandstone conformably.[5] At this time, the Old Red Sandstone was included as part of the base of the Carboniferous System and the Devonian System had not yet been recognized.[11] Sedgwick, on the other hand, started working in north Wales and worked his way south. By 1835, they had finished their work and were proud of what they had accomplished. Murchinson assigned his rocks to the "Silurian System"[12] and Sedgwick his rocks to the "Cambrian System."[13] Further study showed that the lower part of Murchinson's Silurian was identical to the upper part of Sedgwick's Cambrian. Given the amount of work and time they had both invested, neither side was willing to reassign their geologic terrain to the other. Murchinson, however, insisted on extending his Silurian System downward into Sedgwick's unfossiliferous Cambrian rocks. The Cambrian did, however, eventually yield its own distinctive trilobite fossils (*Paradoxides* and *Olenus*).[10] The result was a bitter long-lasting dispute between two friends. After their deaths, as mentioned already, Lapworth employed a solution by assigning the overlapping strata to a new intervening period: the Ordovician.[10] Lapworth, in his 1879 paper, was highly critical of Murchinson's treatment of Sedgwick, but because he mixed in higher social circles, Murchinson's interpretation was favored by the scientific elite of the time; in addition, Murchinson later became head of the British Geologic Survey. The Lower Paleozoic comprised of Cambrian, Ordovician, and Silurian Systems therefore had to wait until near the end of the nineteenth century to be completed. In the meantime, much progress was made by others in identifying younger geologic periods.

The Geology of England and Wales published in 1822 by Conybeare and Phillips contained a large-scale geologic map and stratigraphic column accompanied by very detailed descriptions of the various geologic formations.[13] These authors defined a formation "as a series of beds such as coal, sandstone and shale repeated over and over ... characterized by repetition of its own peculiar members." They resolved the Carboniferous "order" (now called a System) into four formations, from youngest to oldest: the Coal Measures, Millstone Grit, Carboniferous (or Mountain) Limestone, and Old Red Sandstone. Today the Coal Measures and Millstone Grit are assigned to the Upper Carboniferous, and the Mountain Limestone to the Lower Carboniferous. In the United States, these are referred to the Pennsylvanian and Mississippian sub-periods, respectively.

It was subsequently recognized that the Old Red Sandstone, which was deposited in a continental environment, belonged to an older geologic period, the Devonian – below the Carboniferous of Conybeare and Phillips, but above the Silurian of Murchinson. The recognition of the Devonian Period is recounted in great detail in *The Great Devonian Controversy* by Martin Rudwick.[14] Prior to their falling out, Sedgwick and Murchison together visited Devonshire in southern England in 1836 to settle a dispute caused by a claim made by the English geologist Henry De la Beche (1796–1855). De la Beche was doing fieldwork in Devonshire on behalf of the British government, and he reported to the Geological Society of London that he had collected plant fossils identical to those from the Coal Measures in strata much older than the Carboniferous. This finding, if correct, would have had major economic implications since it was thought that economic coal deposits did not exist below the Carboniferous. When Sedgwick and Murchinson arrived in Devonshire to examine the area, they recognized De la Beche had made a serious mistake. He had mistaken the limb of an anticline as a syncline, and he inferred that the fossiliferous rocks were much older than Carboniferous. Further study of the underlying fossiliferous rocks turned out not to be Silurian in age, but transitional with the overlying Carboniferous. Sedgwick and Murchison thus proposed that De la Beche's fossils belonged to a new period in earth history, namely the Devonian.[15]

Subsequent work on the continent recognized less-deformed, equivalent-age rocks in Belgium and the Rhineland, and the Devonian was also recognized in Russia and in the United States. The Devonian marine rocks of Devonshire and the continental Old Red Sandstone of Scotland, for example, were very different lithologies, and confronted geologists with one of the first examples of different "facies" or rocks of the same age deposited under different environmental conditions. The famous unconformity recognized much earlier by Hutton at Siccar Point, Scotland, consisted of Old Red Sandstone unconformably overlying vertical Silurian strata. Hutton, of course, was not aware of their stratigraphic position within the geologic column during his lifetime. The youngest period of the Paleozoic era awaited to be discovered, namely the Permian System.

During the Permian, a shallow continental seaway, the Zechstein Sea, lay over northern Germany and extended into England. The saline conditions produced an impoverished fossil record, making the Permian rocks of Europe difficult to correlate to places elsewhere. A distinctive dolomitic (magnesian) limestone was deposited in Germany and England followed by salt beds and the New Red Sandstone (or red marls), similar to the Old Red Sandstone of Devonian age and the strata described by Conybeare and Phillips.[11] It was not until 1841, however, that Murchinson, after touring Russia, named the Permian System after the town of Perm on the western flanks of the Ural Mountains.[16] These strata were marine and abundantly fossiliferous but Murchison recognized they correlated with the poorly fossiliferous lithologies of the German and English Zechstein. The Paleozoic ends with one of the largest mass extinctions of the Phanerozoic, as the supercontinent Pangaea became assembled and eliminated much of the shallow continental shelf where most marine species lived.

The Mesozoic begins with the Triassic System, which was established by Frederich von Alberti in 1834 in Germany with a three-fold division from bottom to top: the Bunter sandstone, Muschelkalk (shelly limestone), and Keuper (varicolored marl).[17] Only the middle formation represents a shallow-marine deposit transgressing onto the continent, the others being entirely continental. In the Mediterranean region to the south, the Triassic is represented by a thick series of deep sea sediments, deposited in what is today called the paleo-Tethys Ocean, which would in time produce the Alpine mountain range by continental collision. After the Permian mass extinction, the flora and fauna of the Triassic gradually recovered.[3]

The Jurassic was first recognized in the Alps of northern Switzerland in 1799 by Alexander von Humboldt, the most famous of Werner's students, where he applied the term "Jura Kalkstein" (Jura limestone).[2] Alexander Brongniart, coauthor with Cuvier on the Paris basin stratigraphy (see Chapter 1), correlated these strata with the Oolite Series of Britain described by Conybeare and Phillips and also displayed on Smith's 1815 map as a wide southwest-trending swath of Jurassic rocks from Yorkshire to the south coast.[17] The Jurassic is probably best known for its wealth of ammonite fossils, an extinct class of coiled mollusk that proved to be very useful in biostratigraphic correlations on a global scale.[4]

The Mesozoic ends with the Cretaceous System, which was recognized by d'Omalius d'Halloy in 1822 in the Anglo-Paris-Belgian basin and he called *Terrain crétacé*. On his earlier 1815 map, William Smith had recognized four strata blow the Tertiary London clay, namely white chalk, brown or grey chalk, greensand, and micaceous clay (brick earth). Conybeare and Phillips (1822) disagreed with Smith's designation regarding the last formation, and recognized the following two-fold division of the Cretaceous: white chalk, underlain by chalk marl, greensand, and ferruginous sand. They describe in great detail the well-known Cretaceous white chalk with flint nodules in the cliffs of Dover and on the Isle of Wight in southern England.

The subdivision of the Tertiary Period into epochs – namely the Eocene, Miocene and Pliocene – was based on the percentage of extant versus extinct mollusks, and is described by Lyell in his third volume of *Principles* (1833). Referring to marine and fresh-water shell fossil species, Lyell referred to them as: "[T]hese molluscous animals are the medals which nature has chiefly selected to record the history of the former changes of the globe."[6] When Lyell pays tribute to the arduous work of paleontologists in classifying the thousands of different species of mollusks on which he based his stratigraphy, he was likely paying tribute to Gerard Deshayes (1797–1875), the French paleontologist who specialized in Tertiary mollusks. Lyell recognized that his stratigraphy was probably incomplete as the proportion of extinct versus living species, while they increased with age of the strata, varied abruptly rather than smoothly. Subsequently, geologists indeed inserted the Paleocene and Oligocene epochs before and after his Eocene. More recently, the Tertiary was divided into two periods: the Paleogene and the younger Neogene.[2] This completed the Phanerozoic geologic column. A new discovery in physics paved the way to calibrate the age of the geologic record.

RADIOACTIVITY AND RADIOMETRIC DATING

French physicist Henri Becquerel (1852–1908) and the Curies (wife and husband, Marie and Pierre) working in the Sorbonne, Paris, shared the 1903 Nobel prize in physics for the discovery of radioactivity, which had a profound and long-lasting influence on geology as a science.[18] Marie Curie, as part of her doctoral dissertation, discovered radium – a new, highly radioactive element. The New Zealand–born physicist Ernest Rutherford later showed in 1906, working at McGill University, Montreal, that the radiation emitted from radium, called alpha particles, were equivalent to helium nuclei.[19] Rutherford also showed that radium was in a long series of decay products of uranium. He further showed that the ratio of uranium to radium was constant after equilibrium was reached.[19] The final product in this series eventually turned out to be lead, with a total of eight alpha particles (or helium nuclei) being produced in the process:[20]

$$U \rightarrow 8He + Pb$$

These eight heavy helium nuclei produced a substantial amount of heat as they traveled through the surrounding crystal and rock, so much so that the heat could be measured simply with a thermometer and a calorimeter.[21]

Further study showed that the process of radioactive decay was a statistical process for individual radioactive atoms, and that radioactivity obeyed a simple law on the macroscopic scale. Rutherford was again at the forefront of this new research. A student of his at McGill University in 1900, Fredrick Soddy, worked on the radioactive decay of thorium compounds, and together they published their law of radioactive decay in 1902, which stated that the number of atoms that decay depends only on the number of radioactive atoms present:[22]

$$-\frac{dN}{dt} = \lambda N,$$

where N is the number of radioactive atoms present and lambda (λ) is the decay constant which represents the rate of decay (in seconds or years) measured in the laboratory; t is time. Initially, with a large number of radioactive atoms present, the decay rate is rapid, but as the number of radioactive atoms decreases, the rate of decay also decreases (Figure 2.3).

Radioactive decay is analogous to water leaking out of a hole in the bottom of a bucket full of water: the rate of leakage falls in half as the water level reaches half, and slows down accordingly as the water level falls farther. The size of the hole is analogous to the decay constant λ – a large hole corresponds to rapid decay and a small hole to slow decay. If three buckets are connected to each other with different size tubes, eventually the level of water in the second bucket will have a constant ratio to that in the first bucket. In the case of uranium decaying to radium and radium itself also decaying, the ratio of radium to uranium reaches a constant value at equilibrium. In a long decay series, such as that of uranium eventually producing lead at equilibrium, the ratio of lead to uranium depends on time only. As we will see below, this is an important result for radiometric dating of minerals using the ratio of lead to uranium (Pb/U).[18]

At the same time that geologists were rethinking the thermal history of the Earth due to the heat provided by radioactive decay,[20] physicists were following up on Rutherford's original suggestion that minerals and their enclosing geologic formations could be assigned ages using some of the decay products of uranium – namely helium and lead. Rutherford made the first radiometric age estimate of a mineral (fergusonite, a rare earth oxide), yielding an age of 400 million years by the helium/uranium method,[19] which John Joly later recalculated as about 240 million years using newer decay data.[20]

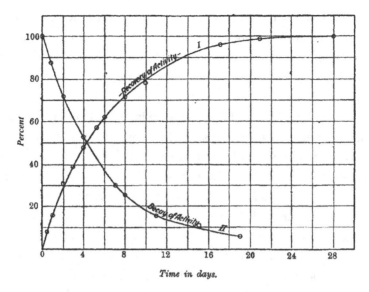

FIGURE 2.3 The law of radioactive decay based on measurements on the radioactive activity of thorium compounds and the daughter product Th.X, now known to be radon (^{224}Ra; see Figure 3.2). The ordinate shows the number of atoms of the parent or daughter, and the abscissa is time. Curve I shows the accumulation of daughter atoms (radon) from thorium decay, and curve II shows the corresponding decay of radon. After about four days, the daughter has declined by half, corresponding to the half-life of radon. Rutherford and Soddy (1902) formulated the law of radioactive decay that says the rate of decay depends only on the number of atoms of the radioactive parent.[22]

Source: Rutherford, E. and Soddy, F. 1902. The radioactivity of thorium compounds. II. The cause and nature of radioactivity. *Journal of Chemical Society of London*, v. 81, 837–860.

The principle of the method was simple but required several assumptions to hold. Helium could not have been added or lost since the mineral's formation. Also, there was no other radioactive source of helium, and the rate of decay of uranium was known. Actually, because the ratio of radium to uranium was known to be a constant, the rate of decay of radium was used instead. Similarly, the Pb/U ratio in a mineral could also be used to provide an age of the same mineral. Since helium is a gas, it was more likely to escape compared to lead, and an early study confirmed this suspicion. R. J. Strutt (later Lord Rayleigh) of Imperial College, London, who was also a mentor to

Arthur Holmes as a student at Imperial, measured the helium/radium ratios in phosphatic nodules (sediments that are rich in radium) with a range of stratigraphic ages. He found poor agreement between calculated ages and stratigraphic level, indicating the unreliability of the helium/uranium method.[23]

In 1905, Bertram Boltwood, working at Yale University, speculated correctly that lead was the final disintegration product in the decay of uranium, and in 1907 he calculated the age of selected minerals from published Pb/U ratios from a variety of locations.[24] The ages ranged from 410 to 2200 million years, but the geologic context of the samples was not well-known, so their value in calibrating the geologic timescale was poor.

Subsequently, Arthur Holmes reexamined the localities from which these samples came and placed the samples into a firmer geologic context.[18] His results are shown in Table 2.3, which compares uranium-helium ages with those of uranium-lead (which tend to be older). This is one of the first attempts to calibrate the geologic timescale. Holmes used the amount of lead produced by uranium decay to recalculate Pb/U ages, and his ages differed somewhat from those of Boltwood who used radium decay to calculate his Pb/U ages. The most commonly dated mineral at the time was uraninite (or pitchblende, a uranium oxide) that typically occurred in pegmatite associated with intrusive granite. Holmes was only twenty-one years old when he published this work, and he spent much of his career improving the calibration of the geologic timescale. He published a new and revised edition of his book *The Age of the Earth* in 1937 and in it he updated the geologic timescale.[25]

In 1947, Holmes compiled the worldwide thickness of sediments for the Phanerozoic and plotted the uranium-lead ages on the abscissa against sediment thickness on the vertical axis, allowing the length of each time period to be estimated.[26] This timescale, however, later needed revision to extend it to longer time periods.[27] Part of the problem was that many of the mineral ages were from complex geologic terrains. The Appalachian mountain belt, for example, records

Table 2.3 *Ages for geologic systems (Holmes 1913)*[18]

Geologic Systems	Timescale in Millions of Years	
	Helium Ratio	Lead Ratio
Pleistocene	1	
Pliocene	2.5	
Miocene	6.3	
Oligocene	8.4	
Eocene	30.8	
Cretaceous		
Jurassic		
Triassic		
Permian		
Carboniferous	146	340
Devonian	145	370
Silurian		430
Ordovician	209	
Cambrian		
Algonkian		1,000–1,200
Archean		1,400–1,600

Note: U/Pb data from Boltwood (1907).[24]

evidence for three mountain-building events: the Taconian (Ordovician), the Acadian (Devonian), and the Alleghanian (Permian). Minerals that gave Devonian ages were actually formed during the Ordovician event, and minerals that yielded Permian ages were actually formed during the Devonian event, having lost daughter products during the subsequent Permian metamorphism. The resulting timescale was therefore artificially compressed.

By the 1960s, however, the geologic timescale was beginning to look somewhat similar to the modern timescale (Figure 2.4). One of the reasons for this is that in addition to the U–Pb systems of dating, two new radiometric methods were developed: the potassium–argon method (K/Ar), where potassium-40 is the parent and argon-40 is the daughter, and the rubidium–strontium method (Rb/Sr), where

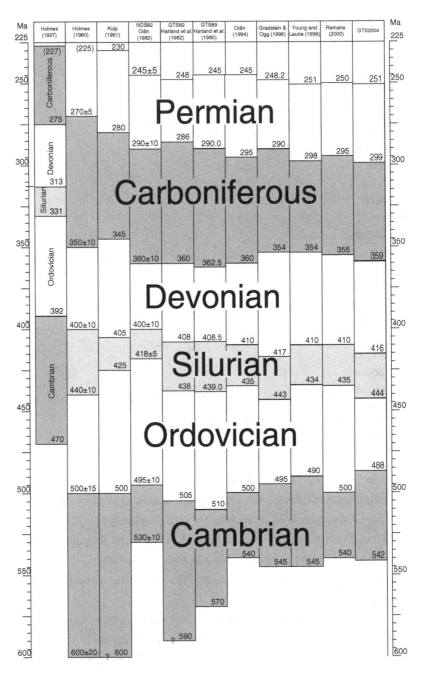

FIGURE 2.4 Evolution of part of the Phanerozoic geologic from Cambrian to Permian during the twentieth century. Holmes' 1937 and 1960 timescales are shown on the left hand side. The right hand side shows the 2004 timescale from Gradstein and others. Since the 1960s the changes have been minor with the exception of the base of the Cambrian, now placed at 542 million years.

Source: Gradstein, F., Ogg, J. and Smith, W. 2004. *A Geologic Time Scale.* Cambridge University Press, Cambridge.

rubidium-87 is the parent isotope and strontium-87 is the daughter (see Chapter 8). These systems could be applied to common potassium-rich minerals such as the micas and alkali feldspars. The second reason the geologic timescale improved around this time is that loss of daughter products (such as argon, lead, or strontium) during metamorphism became better understood, and ages could be cross-checked by different dating methods. Figure 2.4 shows the evolution of the timescale over time where it can be seen that Holmes's 1960 and Kulp's 1961[28] timescales are similar to each other, and subsequent revisions are relatively minor (with the exception of the base of the Cambrian, which is now placed at 542 million years ago).

DINOSAUR EXTINCTION AT THE CRETACEOUS–PALEOGENE BOUNDARY

There were five major extinctions during the Phanerozoic, with an average of one extinction every 100 million years.[29] The boundary between the Mesozoic Era (middle-life) and Cenozoic Era (new-life) also corresponds to the boundary between the Cretaceous (Latin for chalk, *creta*) and Tertiary periods. The Cretaceous-Tertiary boundary is one of the most studied geologic boundaries because, as all geologists and most seven-year-olds know, it corresponds to the extinction of the dinosaurs, 65 million years ago. It also corresponds to the extinction of a large percentage of both marine and terrestrial animals and plants. The ammonoids and belemnites, both mollusks that were used to sub-divide the Cretaceous biostratigraphically, also became extinct at the end of the Cretaceous, as did swimming reptiles and most species of plankton.[29] In recent geologic timescales, the term Tertiary is regarded as antiquated (belonging with the older terms Primary and Secondary) and is replaced by two new periods, the older Paleogene and the younger Neogene. Today the Cretaceous-Tertiary boundary is referred to as the Cretaceous–Paleogene (K–Pg) boundary.

In 1980, an important paper was published by the father-and-son team Luis and Walter Alvarez (Luis a Nobel Prize–winning physicist, and Walter a geologist), and also two chemists, Frank Asaro and Helen Michel, who proposed that a large meteorite impact caused the K–Pg extinction.[30] They found a spike of iridium (a member of the platinum group of elements) in the clay layer at the boundary in Italy, Denmark, and also New Zealand, and this spike has since been found by other workers in dozens of localities around the world. Since iridium is extremely rare in the Earth's crust (about one part in ten billion) but is much more abundant in meteorites, the call for an extraterrestrial source of the iridium was justified.

The authors explain in their paper how they happened to measure iridium at the K–Pg boundary. Previous work had shown that the iridium concentration in deep sea sediments depended on the sedimentation rate, and as Walter Alvarez was trying to estimate the timescale for the K–Pg boundary, he measured iridium concentrations. When the spike was found, Walter informed his father, who knew that meteorites had a relatively high iridium content. Thus, Luis Alvarez suggested the impact hypothesis.

Also found at the K–Pg boundary worldwide was other evidence of a meteorite impact, including glass spherules, shocked crystals of quartz, and a high-pressure form of quartz, stishovite.[30] The glass spherules are produced by target rocks that are melted on impact and thrown into the atmosphere, and then rapidly cool and settle back to earth as glass spheres. Shocked quartz is produced when a shock wave passes through the rock and is only found at nuclear bomb test sites and meteorite impact sites, and it is not produced by violent volcanic eruptions. Similarly, stishovite is only produced by nuclear explosions and meteorite impacts. In order to produce the observed iridium anomaly, the size of the putative meteorite is estimated to be 10 kilometers in diameter, and it should have produced a crater about 150 kilometers in diameter based on laboratory experiments. Without identifying the actual site of impact, however, many scientists remained skeptical of the impact hypothesis. Alvarez and others noted at the end of their paper that the impact site may have been subducted through plate tectonics, never to be found. It is interesting to note that experts consider that a 150 kilometers wide meteorite crater occurs on Earth approximately every 100 million years, leading to speculation that the other four major extinctions were also caused by meteorite impacts – but the geologic evidence does not bear this out.[29]

Eleven years after the Alvarez team's 1980 paper, a group of scientists published a paper on a possible Cretaceous–Paleogene impact crater on the Yucatán Peninsula, Mexico.[31] This circular structure was about 180 kilometers in diameter, defined by a gravity anomaly similar to that of known impact structures, and also has a magnetic anomaly. This structure had already been drilled by a Mexican oil company, so the stratigraphy was already known. It consisted of carbonate and sulfate continental shelf rocks underlain by intermediate igneous rocks of andesitic composition. In several drill-hole cores, thick breccias and melted andesite glass were encountered. These rocks were similar in composition to impact melt rocks in nearby Haiti of the same age. Shocked quartz was found in two of the wells, but were not present in wells outside the structure. Perhaps the best evidence that this structure was the culprit is that nearby sites at the K–Pg boundary in the Gulf of Mexico and the Caribbean showed the greatest thickness of impact rocks, and the overall evidence for impact progressively decreased away from these sites in the western interior of North America and other more distant, global sites. This structure appeared to be the "smoking gun" to confirm the Alvarez impact hypothesis.[32]

The still more difficult question is what exactly killed the dinosaurs? The potential catastrophic environmental effects of such a large impact are many. Originally the 1980 paper suggested – by analogy with the ash sent aloft by the largest volcanic explosion on record in Krakatoa, Indonesia, in 1883 – that the impact caused sunlight to be blocked out by the dust injected into the atmosphere. Day became night for several years, causing photosynthesis to shutdown, which in turn caused the demise of dinosaurs that depended on vegetation and, thus, also their predators. Now that the target rocks on the Yucatán Peninsula are known to be carbonates (limestone) and sulfates (anhydrite), other environmental catastrophes present themselves. Vaporization of the carbonates could have produced enough CO_2 to cause a 10°C global warming that lasted in excess of 10,000 years. Alternatively, vaporization of sulfates would produce sulfur aerosols, which at first reflect sunlight and causes cooling, but then will precipitate as acid rain thereby stressing the vegetation. Such acidification could also have destroyed shallow marine life, but probably not life in the deep ocean. A global conflagration in the form of wildfires has also been suggested, but a carbon-rich soot layer has only been found at a few K–Pg sites. Most of these scenarios are based on theoretical models and the data are equivocal, but it is quite possible that several of these mechanisms (acidification, darkness, greenhouse gases, a toxic atmosphere) operated in concert to cause the observed extinctions, in both terrestrial and marine environments. The K–Pg impact hypothesis is a fascinating and well-documented example of how the scientific process works. The entire story is told in detail in *The End of the Dinosaurs* by Charles Frankel.[33]

SUMMARY

The geologic column was built up by painstaking and arduous field-work, largely in the first half of the nineteenth century. Murchinson, who is credited with discovering three of the geologic periods (Silurian, Devonian, and Permian), traveled 13,000 miles alone on his 1840 Russian trip, and William Smith likely covered more mileage in order to produce his 1815 map of England and Wales. The economic importance of such maps was recognized by governments who supported the mapping efforts, but the geologic column appears to have been assembled piecemeal by gentleman specialists as independent workers in different countries. It was only after Smith completed his map that he received a government pension for his efforts.

It was the suggestion of physicist Ernest Rutherford in 1906, after the discovery of radioactivity, that geologic formations could be dated by minerals containing uranium allowing calibration of the geologic timescale to begin. The first mineral ages were either helium-uranium or lead-uranium ages, and are now only of historical interest. Arthur Holmes played a major role in improving calibration of the geologic timescale over a fifty-year period from 1911 to 1960. The geologic timescale is still a work in progress.

One of the most studied geologic boundaries is the Cretaceous-Tertiary (now called the Cretaceous–Paleogene) boundary, which corresponds to one of the major Phanerozoic extinction events, including that of the dinosaurs. Consensus has largely been reached that it was caused by a meteorite impact on the Yucatán Peninsula 65 million years ago. The next chapter addresses the early efforts to date the age of the Earth.

REFERENCES

1. McPhee, J. 1982. *Basin and Range*. Farrar, Straus and Giroux, New York.
2. Harland, W. B. et al. 1982. *A Geologic Time Scale*. Cambridge University Press, Cambridge.
3. Gradstein, F., Ogg, J., and Smith, W. 2004. *A Geologic Time Scale*. Cambridge University Press, Cambridge.
4. Gradstein, F., Ogg, J., and Smith, A. 2012. *The Geologic Time Scale*. Cambridge University Press, Cambridge.
5. Geikie, A. 1905. *The Founders of Geology*, 2nd ed. Macmillan, London.
6. Lyell, C. *Principles of Geology*. 1880–1883 (1997). Penguin Classics, London.
7. Steno, N. 1669. De solido intra solidum naturaliter content dissertattionis prodromus. Florence. Private publisher.
8. Strachey, J. 1717. A curious description of the strata observed in the coal-mines of Mendip, Somersetshire. *Philosophical Transactions Royal Society London*, v. 30, 968–974.
9. Phillips, J. 1841. *Figures and Descriptions of the Palaeozoic Fossils of Cornwall, Devon and West Somserset*. Longman, Brown, Green and Longmans, London.
10. Lapworth, C. 1879. On the tripartite classification of the Lower Paleozoic rocks. *Geology Magazine*, v. 6, 1–15.
11. Conybeare, W. D. and Phillips, W. 1822. *Geology of England and Wales*. Phillips, London.
12. Murchinson, R. 1852. On the meaning of the term "Silurian System." *Proceedings of Geological Society of London*, v. 8, 173–181.

13. Sedgwick, A. and Murchinson, R. 1836. On the Silurian and Cambrian Systems, exhibiting the order in which older sedimentary strata succeed each other in England and Wales. *British Association for Advancement of Science Report*, 5th meeting 1835, Dublin, 59–61.

14. Rudwick, M. 1985. *The Great Devonian Controversy.* Chicago University Press, Chicago.

15. Sedgwick, A. and Murchinson, R. 1839. On the older rocks of Devonshire and Cornwall. *Geological Society of London Proceedings*, v. 3, 121–123.

16. Murchinson, R. 1841. First sketch of some of the principal results of a second geological survey of Russia, in a letter to M. Fischer. *Philosophical Magazine*, v. 19, 419–422.

17. Von Zittel, K. 1901. *History of Geology and Palaeontology.* Translated by M. Ogilvie-Gordon. Scott, London.

18. Holmes, A. 1913. *The Age of the Earth.* Harper & Brothers, London.

19. Rutherford, E. 1906. Mass and velocity of the alpha particles expelled from radium and actinium. *Philosophical Magazine*, v. 12, 348–371.

20. Joly, J. 1909. *Radioactivity and Geology.* Constable & Co, London.

21. Curie, P. and Laborde, A. 1903. Sur la chaleur dégagé spontanément par sels de radium. *Comptes Rendus de Séances de l'Academie de Sciences*, v. 86, 837–860.

22. Rutherford, E. and Soddy, F. 1902. The radioactivity of thorium compounds. II. The cause and nature of radioactivity. *Journal of Chemical Society of London*, v. 81, 837–860.

23. Strutt, R. J. 1908. On the accumulation of helium in geologic time. *Proceedings of the Royal Society of London, A*, v. 81, 272–277.

24. Boltwood, B. 1907. On the ultimate disintegration products of radioactive elements. Part II. The disintegration of uranium. *American Journal of Science*, v. 23, 77–88.

25. Holmes, A. 1937. *The Age of the Earth.* Nelson & Sons, London.

26. Holmes, A. 1947. The construction of a geological time-scale. *Transactions of Geological Society of Glasgow*, v. 21, 117–152.

27. Holmes, A. 1960. A revised geological time-scale. *Transactions of the Royal Society of Edinburgh*, v. 17, 183–216.

28. Kulp, J. 1961. Geologic time scale. *Science*, v. 133, 1105–1114.

29. Raup, D. M. 1992. *Extinction.* Norton & Co, New York.

30. Alvarez, L. W., Alvarez, W., Asaro, F., and Michel, H. 1980. Extraterrestrial cause of the Cretaceous-Tertiary extinction. *Science*, v. 208, 1095–1108.

31. Hildebrand, A. R. et al. 1991. Chicxulub crater: a possible Cretaceous/Tertiary boundary impact crater on the Yucatan Peninsula, Mexico. *Geology*, v. 19, 867–871.

32. Schulte, P. et al. 2010. The Chicxulub asteroid impact and the mass extinction at the Cretaceous-Paleogene boundary. *Science*, v. 327, 1214–1218.

33. Frankel, C. 1999. *The End of the Dinosaurs*. Cambridge University Press, Cambridge.

34. Davies, G. H. 1981. The history of the Earth sciences. In *The Cambridge Encyclopedia of Earth Sciences*. Cambridge University Press, Cambridge.

3   A Vestige of a Beginning

The Age of the Earth

> It must indeed be admitted that many geological writers of the
> "Uniformitarian" school, who in other respects have taken a profoundly
> philosophical view of their subject, have argued in a most fallacious
> manner against hypotheses of violent action in the past.

– Lord Kelvin (1864)[1]

INTRODUCTION

By the time Charles Lyell's well-received and popular book *Principles* was published in the 1830s, Neptunism had largely disappeared as a viable geologic theory.[2] However, shortly thereafter, a vigorous new controversy arose between the renowned physicist and mathematician Lord Kelvin (William Thompson) and Lyell's followers, including Charles Darwin. These eminent geologists and biologists believed, as did James Hutton, that the processes operative today could explain the evolution of the Earth without invoking more intense or catastrophic events – so long as sufficient time was available. Many of Hutton's followers were accused of admitting infinite time in part based on the last sentence of his 1788 paper: "There is no vestige of a beginning and no prospect of an end."[3]

In his paper "On the Secular Cooling of the Earth" (1864), Kelvin used Fourier's laws of heat conduction to calculate that the Earth was between 20 million and 400 million years old; the large age range was due to lack of information on the thermal properties of rocks at depth in the Earth at that time.[1] Kelvin had earlier calculated the age of the Sun, assuming (incorrectly) that its energy came from gravitational accretion, and arrived at an age of about 20 million years.[4]

The Sun's energy actually comes from nuclear fusion, but that would not be recognized until the 1930s.

The agreement between Kelvin's age of the Sun and the Earth (at the lower age range) was pure coincidence, but produced one of science's most heated controversies of the nineteenth century. Kelvin's estimate for the age of the Earth (and the Sun) was based on elegant mathematics and delivered by the most authoritative physicist of his day, and had a profound impact on geology. Now geologists had two big problems – not enough time to account for the sedimentary record, and that Lyell's doctrine of uniformity contradicted the second law of thermodynamics.

KELVIN AND THE SECOND LAW

Sir William Thompson (1824–1907), later Lord Kelvin, was born in Belfast, Ireland, to James Thompson, who was an engineering professor at Belfast and later at Glasgow.[5] William's mother died when he was six. William attended Glasgow University at a young age and entered Cambridge in 1841, where he excelled at rowing, graduating in 1845. After a brief visit to Paris to gain laboratory experience and to meet the eminent continental scientists of the day, he returned to Glasgow to take the chair in physics (natural philosophy as it was then called) in 1846. He is best known for his theoretical development of the absolute temperature scale (Kelvin scale), and his statement of the second law of thermodynamics, published in 1851 under the title "On the Dynamical Theory of Heat." He was also a practical engineer who made his fortune in telegraphy and submarine cable communication. He was an unusually productive scientist and published over 600 papers, achieved many patents, and received numerous of the highest scientific awards in England at the time. He is regarded as the founder of physics in Britain.

Kelvin's disagreement with Lyell's uniformitarian school in the latter half of the nineteenth century was just as much about how the second law of thermodynamics was inconsistent with the uniformity of geological processes as it was about the age of the Earth. The word thermodynamics refers to heat (*thermo*) and work (*dynamics*).

The first part of Kelvin's 1864 paper points out that Fourier's "beautiful" solutions to the heat-flow laws showed that heat dissipates and temperatures become uniform with time. One of the statements of the second law of thermodynamics, due to the French scientist Rudolf Clausius (1822–1888), is that heat flows from hot to cold bodies. Based on this statement, the Earth must be losing heat because the temperature in the Earth increases with depth; thus, heat flow must be outward into space, assuming there is no chemical internal heat source, which Kelvin thought unlikely. Therefore, plutonic and volcanic activity of the past must have been more active rather than being uniform throughout time. In the 1864 paper, Kelvin takes Charles Lyell to task for explaining underground heat and plutonic action by a chemical hypothesis:

> That the substances, combining together, may be again separated electrolytically by thermo-chemical currents due to the heat generated by their combination, and thus the chemical action and its heat continued in an endless cycle, violates the principles of natural philosophy.[1]

Further on, Kelvin has some fun at Lyell's expense by saying that this was the same as how "a clock constructed with a self-winding movement may fulfill the expectations of its ingenious inventor by going on forever."[1] In other words, there is no such thing as a mechanical or electrical perpetual motion machine. Thirty-three years later, Kelvin repeated this same criticism verbatim in his second 1897 paper on the age of the Earth, indicating that, in his view, little had changed amongst geologists.[6] Kelvin formulated his own statement of the second law of thermodynamics (equivalent to that made by Clausius), namely: it is not possible to absorb heat from a hot reservoir and convert it entirely to work without losing some of the heat to a colder reservoir (for example, the environment). This statement of the law makes clear that perpetual motion machines are impossible because some of the energy is always wasted. Although the first law of thermodynamics says that heat and work are equivalent and can be converted from one to the other, the second law imposes a caveat or restriction on the first law limiting the conversion of heat to work. That radioactivity (though not yet discovered at the time of these arguments) provided an internal heat source for the Earth does not invalidate the second law, but it does change the quantitative analysis very considerably.

KELVIN AND THE AGE OF THE EARTH

While the first part of Kelvin's important 1864 paper pointed out the inconsistency of the uniformitarian viewpoint with the second law of thermodynamics, the second part focused on Fourier's solution to the heat-flow equation to calculate the age of the Earth. Following Leibniz's earlier theory that the Earth was initially a hot incandescent sphere that cooled to form a crystalline crust, Kelvin calculated the time necessary for the Earth's crust to reach its current geothermal gradient.[7] As the Earth cooled, the geothermal gradient would gradually decrease to reach its current value; Kelvin estimated the time required for this to occur, assuming no internal heat source.

Kelvin needed three quantities for his calculation: the initial temperature of the Earth; the thermal conductivity and heat capacity of the rocks, and how they might vary as a function of depth; and the geothermal gradient. For the geothermal gradient, he chose 1°F per 50 feet (about 35°C/km) based on measurements for the continental crust (this figure is not applicable to the ocean crust, of which little was known at the time). He made his own measurements earlier on the thermal conductivity of rocks and used an average value of 400 ft^2/year (1.2 x 10^{-6} m^2/sec) which is still used by modern researchers for average rock. His estimate for the initial temperature of the Earth was 7,000°F (3,870°C), which was much too high. The potentially large range in these variables led Kelvin to take a conservative approach, and he ultimately concluded that the Earth was within the wide range of 20 million to 400 million years old. As noted already, the lower limit also coincided with his earlier estimate for the age of the sun. Clearly the Earth cannot be older than the sun, so this placed a maximum age of 20 million years on the Earth. In his *Origin of the*

Species (1859), Darwin had estimated that denudation of formations in southern England required 300 million years, but he later removed this calculation from his second edition.[8] Kelvin's work had put both geologists and biologists in a bind as Darwin also needed lots of time for his own theory of natural selection.

A second estimate for the age of the Earth was made by an American geologist. The United States Geological Survey was formed in 1878, and its first director was Clarence King, born in Newport, Rhode Island, in 1842. His tenure at the Geological Survey was short as administrative work was not to his liking; he left the survey three years later in 1881 and turned to mining ventures and consulting work.[9] He was, however, a well-respected geologist and was elected to the National Academy of Sciences in 1876. While at the Survey, he set up a geophysical laboratory to make pressure and temperature measurements on rock samples.

King apparently had a very specific plan. He was interested in constraining the wide age range Kelvin proposed for the Earth in his 1864 paper, but King knew he needed additional data on the melting behavior of rocks at depth in the Earth to make further progress. He hired a new Ph.D., Carl Barus, to do the rock experiments, and they agreed that Barus would publish the results independently while King would interpret the data's geological significance. Barus published data on the melting properties of diabase, a rock of basaltic composition, thought to be representative of the Earth's interior at the time. He ended one of his published data contributions with the following comment: "The immediate bearing of all this on Mr. Clarence King's geological hypothesis is now ripe for enunciation."[10]

King indeed followed up with a paper entitled "The Age of the Earth," using Barus's new data in 1893.[11] His point of departure in the paper was the conclusion of Kelvin, and others before him, that the Earth has a rigidity as high as steel, thereby precluding a large amount of liquid at depth. He used the data of Barus on the melting behavior of diabase, which has a melting temperature of 1,200°C at the surface but increases to about 4,000°C at depth, according to Barus's calculations. King showed that Kelvin's 100-million-year cooling curve lay on the liquid side of the diabase melting curve, which was inconsistent with a highly rigid Earth. He concluded that 24 million years was a reasonable maximum age for the Earth since this curve fell on the solid side of the diabase melting curve. Kelvin, in his second paper on the age of the Earth, agreed with King's calculation, and Kelvin concluded, "I am not led to differ much from his [King's] estimate of 24,000,000 years ... it would be quite inadvisable to publish any closer estimate."[6] Geologists were now in a much tighter bind than before. Instead of a range of 20 million to 400 million years as in the 1864 paper, the maximum age was now at the lower end of this range.

As an illustration of some aspects of Kelvin's (1864) and King's (1893) models, based on Fourier's solution, Figure 3.1 shows different cooling curves for a semi-infinite solid with an initial temperature of 1,200°C (used by King) and a surface temperature of 0°C. Other variables are the same as those used by Kelvin. Cooling curves at 30 million, 13.5 million, and 7.6 million years correspond to near surface geothermal gradients of 20, 30, and 40°C/km, respectively. The geothermal gradients between 20°C/km and 40°C/km correspond to Earth ages of 7.6 to 30 million years, which is consistent with Kelvin's and King's intermediate estimate of circa 24 million years.

AGE OF THE OCEANS

When speaking of the age of the Earth, three different meanings can be recognized – the formation of the planet in an astronomical sense, the age of the first formed crust after cooling from a hot sphere (Kelvin's calculated age), and the age of the oceans. The astronomer Edmund Halley (1656–1742) appears to be the first to suggest that the age of the ocean could be estimated from its salinity and the rate of introduction of salt by rivers into the ocean.[12] Halley's argument utilized an analogy based on lakes that had rivers entering but not exiting the lake, such as lakes with internal drainage, for example the Caspian Sea. The salinity of such lakes should increase with time due to input from the dissolved load of rivers and also evaporation of the water. If the lake

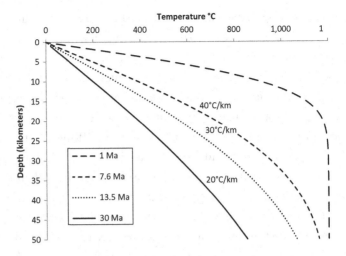

FIGURE 3.1 Cooling curves for the Earth without an internal heat source based on cooling of a semi-infinite solid initially at 1,200°C and 0°C surface temperature. These curves illustrate the calculations done by Kelvin[1] and King[11] in their estimates for the age of the Earth. The geothermal gradient decreases to 40, 30, and 20°C/km after 7.3 million, 13.5 million, and 30 million years of cooling. The latter two times bracket Kelvin's and King's estimate of 24 million years. The discovery of radioactivity invalidated these models and provided geologists with longer times to account for sedimentary accumulations.

Source: Kelvin, Lord (Thompson, W.). 1864. On the Secular Cooling of the Earth. *Transactions Royal. Society Edinburgh*, v. 23, 157–169; Carslaw, H. S. and Jaeger, J. C. 1959. *Heat Conduction in Solids*. Clarendon, Oxford.

waters were sampled several hundred years apart, an increase in salinity should be observed and "by the rule of proportion" estimates, the total time required to "acquire its present degree of saltiness." The age of the oceans could be similarly calculated, and Halley stated, "[I]t were to be wished that the ancient Greeks and Latins had delivered down to us the degree of saltiness of the sea as it were 2000 years ago."[12] Halley realized his suggestion was only hypothetical, and urged that the relevant data on the ocean and lake salinity be collected. He also recognized that if the oceans were initially salty, the age estimate would be a maximum value and that this "would refute the ancient notion that some have of late entertained, of the eternity of all things."[12] He concluded his 1715 paper with the following: "the world may be much older than many have hitherto imagined."[12]

Halley's suggestion appears to have been forgotten until 1876, when T. M. Reade used the chloride and sulfate concentrations of calcium and magnesium in the oceans and the input of these compounds from rivers. He termed his method as one of "chemical denudation."[13] His results ranged from 25 to 200 million years. Because limestones, dolomites, and sulfates are precipitated from the oceans (thereby resetting the salinity clock), these estimates were not thought to be valid by some, including the Irish geologist John Joly.

John Joly (1857–1933) was born in County Offaly, Ireland, and was educated at Trinity College Dublin, where he graduated in 1882 with a single degree in engineering and the physical sciences.[14] He was especially interested in geology. He became a professor of geology and mineralogy at Trinity, a post he kept until his death. He developed several important inventions, including a calorimeter to measure the specific heat of minerals and a method to measure small amounts of the newly discovered radioactive element radium. His book *Radioactivity and Geology* (1909) shows that he was at the forefront of both geology and the recently discovered phenomenon of radioactivity.[15] He is best known for his estimated age of the oceans based on their sodium content, published in 1899.[16]

Joly's method was similar to Reade's, but instead he focused on estimating the total amount of the element sodium in the oceans divided by the amount of sodium in solution contributed annually to the oceans by the world's rivers, utilizing this formula:

$$\text{Age of oceans} = \frac{\text{Total Na in ocean (tons)}}{\text{Na influx from rivers (tons/yr)}}.$$

His determined an estimated age for the oceans of 80 million to 90 million years, but allowed a possible maximum age by this method of 150 million years in his later 1923 paper.[17] By this time he was aware that Pb/U chemical ages were yielding similar and older ages. In keeping with the uniformity doctrine, Joly assumed a uniform chemical dissolution rate by rivers from the land surface, and he also assumed constant land area and rainfall "within wide limits," but with a long-term constant average over geologic time. He chose the element sodium as the sole tracer because of its high solubility as opposed to sulfates and carbonates of elements, such as Ca and Mg, which were precipitated from solution as marine sedimentary deposits.

To insert values into the simple equation above, a substantial amount of information was needed, and Joly relied heavily on reports from the HMS Challenger oceanic expedition (1872–1878), which provided salinity and chemistry of collected water samples. Sir John Murray (1841–1914), who sailed on the Challenger and later organized the publication of the results, provided information on the world's rivers. The quantities needed to complete the equation included the volume and mass of the ocean, its salinity, the total mass of sodium in the ocean, the mass of sodium contributed by the world's rivers, and also the annual river discharge. The initial results produced an age of 90 million years using the following figures:

$$\frac{14{,}151 \times 10^{12}}{15{,}727 \times 10^{4}} = 90 \times 10^{6} \text{ yrs.}$$

A substantial number of variables that could introduce both positive and negative errors on this estimate were discussed at length by Joly, including a highly speculative estimate of the initial salinity of the oceans, the amount of sodium involved in evaporation of ocean water, a coastal erosion contribution, changing paleogeography, different rock types exposed, salt evaporites deposited, and many other variables. Evaluation of these variables was poorly constrained, but Joly's final estimate did not differ substantially from that above. Today, oceanographers recognize that Joly was actually calculating the residence time of Na in the oceans, which is estimated to be about 68 million years.[18]

EARTH'S SAND HOURGLASS

Perhaps spurred on by Kelvin's earlier (1864) paper, many of the best-known geologists of the time tried to quantify their fledgling science by estimating the age of the Earth from the total stratigraphic thickness of the known geologic column. Arthur Holmes, in his book *Age of the Earth* (1913), lists nineteen such attempts between 1860 and 1909.[19] The stratigraphic method is simple in principle; if the thickness of the stratigraphic record is known and the rate of deposition of the strata is estimated or assumed (in years per unit thickness), the time to accumulate the strata is simply:

Time = (sedimentary thickness) × (rate of deposition).

In keeping with Lyell's uniformitarianism, these geologists assumed a constant rate of deposition throughout the geologic column – but they differed amongst each other on this rate. As Arthur Holmes noted, "uniformity proved a great advance, but in detail it is apt to lead us astray if applied too dogmatically."[19] The time needed by these various investigators, omitting outliers, averaged about 100 million years, substantially in excess of Kelvin's and King's geophysical estimates discussed earlier in this chapter. Most of these sedimentary estimates excluded pre-Cambrian rocks, of which little was then known; but as we now know, this period amounts to about 88 percent of all geological time, so these estimates had no chance of being even close to correct. It is interesting to follow the attempts of one geologist to date the Phanerozoic (Cambrian to Recent) sedimentary record.

William Johnson Sollas (1849–1936) was born in Oxford, England, and was educated at the Royal School of Mines, London, and at Cambridge University with degrees in geology.[20] He taught geology as part of the university extension system in various parts of the country. He became professor of geology at three different universities: the Universities of Bristol (1880–1883), Trinity College, Dublin (1883–1897), and then University of Oxford (1897–1936). He was president of the Geological Society of London from 1908 to 1910.

Table 3.1 *Stratigraphic studies by Sollas*[19]

Publication Date	Stratigraphic Thickness ($\times 10^3$ meters)	Deposition Rate (year/meter)	Age (Ma)
1895	50	328	17
1900	81	328	26.5
1909	102	328	80

Sollas had a long and productive career, and died at the age of 87. He was interested in the question of the age of the Earth and published three papers based on the stratigraphic record on the topic. The age estimates derived by Sollas in his publications are summarized in Table 3.1.

As more geologic information became available, the stratigraphic column increased in thickness in the late nineteenth and early twentieth century, but Sollas's estimated age increased by a much greater factor in his 1909 study. In the first two studies, only Cambrian and younger sediments are accounted for. In the most recent study (1909), the age obtained by multiplying column 2 by column 3 (Table 3.1) gives an age of only 33.5 million years. Since this is much lower than the age of the oceans calculated by Joly above (90 million years), there was a serious problem for Sollas since the sediments were almost entirely deposited in the ocean implying a long time period devoid of any oceanic deposition.

Sollas increased his estimate to 80 million years by two means: he included time for pre-Cambrian sediments (he added 17.5 million years after discussing the sparse geologic data), and then he included time for missing strata at unconformities. After discussing the geologic data for six major unconformities, he added 24 million years and for five minor unconformities he added 5 million, thus bringing the total to 80 million years.[21] Since very little was known about pre-Cambrian strata and the missing time represented by unconformities, these additions can only be regarded as mere guesses rather than indicating an agreement between the age of the oceans based on Joly's estimate and the stratigraphic record. Sollas presented these results at his annual address to the Geological Society in 1909. There, his introductory comment places the situation of geologists within their historical context:

> The question of geologic time has ceased to be made a cause of reproach to us, and we no longer are suspected of an overdrawn account in a metaphoric bank of time: indeed, since physics, in the language of the Stock Exchange, has forsaken its role as bear for that of bull, we seem rather to be threatened with the novel embarrassment of having more time on our hands than we know how to dispose of.[21]

Sollas was of course referring to the Kelvin controversy when geologists were only allowed about 25 million years of time, and to the post-radioactivity discovery period where radiometric ages of minerals were now yielding ages well in excess of a hundred millions years only a decade after the discovery of radioactivity in the late 1890s. This newly emerging field of research would eventually lead to the correct estimate of the age of the Earth.

ISOTOPIC AGES

In the previous chapter we saw how simple lead uranium ratios (Pb/U) were used to calculate mineral ages and these are referred to as "chemical ages" and they are distinctly different from isotopic ages (Chapter 8). Isotopes are atoms of the same element, but with a different number of neutrons in the nucleus and their chemical behavior is the same but their physical properties differ. Their existence was suspected by Fredrick Soddy, co-discoverer of the law of radioactive decay, when it was observed that the atomic weights of the elements were not whole integer numbers, suggesting natural elements were a mixture of isotopes of the same element. In the early decades of the twentieth century, physicists established that there are four isotopes of lead, three of which are produced by three separate parent-daughter decay schemes (Figure 3.2):

$$^{238}U \rightarrow {}^{206}Pb \qquad U \rightarrow {}^{207}Pb \qquad Th \rightarrow {}^{208}Pb$$

The superscripts here refer to the atomic weights so that the only difference between ^{207}Pb and ^{206}Pb is one extra neutron in the nucleus. The fourth lead isotope is ^{204}Pb, which does not have a radioactive parent and represents primeval (or initial lead) present at the time the Earth's crust formed. There is good reason, however, to believe that this initial lead is not just ^{204}Pb, but also comprises unknown amounts ^{207}Pb, ^{206}Pb, and ^{208}Pb. Any attempt to date the age of the Earth must be able to distinguish between primeval and radiogenic lead, and this turned out to be a difficult task – the answer would lie with meteorites.

Before the abundance of the lead isotopes could be measured, the mass spectrometers of the 1930s, which only measured the abundances of lighter isotopes with any precision, needed improvement. Physicist Alfred O. C. Nier, working as a postdoctoral fellow at Harvard in 1937 and later at the University of Minnesota, made significant improvements to the mass spectrometer and allowed the first precise measurements of lead isotopes in minerals, mainly galena (PbS), a lead sulfide.[22,23] This mineral does not contain radiogenic lead since it contains little or no uranium or thorium. Nier and his colleagues subsequently measured radiogenic lead isotopes in a variety of minerals of different ages, mainly pitchblende (UO_2) from around the world, in 1939.[24] These minerals contained two types of lead: primeval lead and radiogenic lead due to decay of uranium and thorium in the mineral. These measurements subsequently led a number of workers to independently propose a general model for the evolution of lead isotopes in the Earth and the first U–Pb isotopic ages for the Earth – these workers were Gerling (1942)[25], Holmes (1946[26], 1947[27]) and Houtermans (1946[28]), all of whom used Nier's data and arrived at similar estimates for the age of the Earth: about 3 billion ± one billion years. Their method is commonly called the

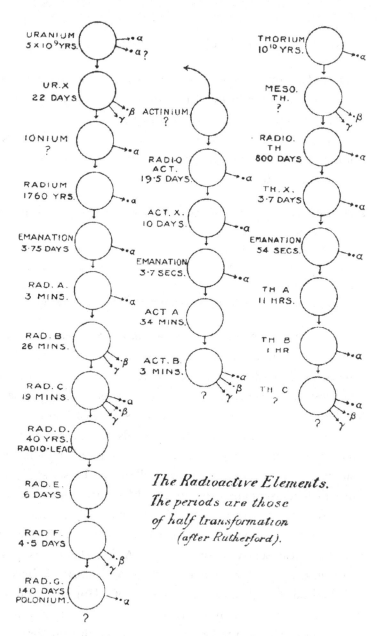

FIGURE 3.2 Three tentative radioactive decay series, with numerous question marks indicated, after E. Rutherford, updated by J. Joly. The first two columns are part of the uranium series (now known to be ^{238}U and ^{235}U, respectively), and the third column is the thorium (^{232}Th) decay series. "Ionium" in the first column is now known to be ^{230}Th. "Emanation" refers to various radioactive decay products such as alpha radiation. UR.x and TH.x in the first and third columns refer to the decay products of ^{238}U and ^{232}Th, namely radon gas. The final stable isotopes in the uranium and thorium series are now known to be ^{207}Pb, ^{206}Pb, and ^{208}Pb. Times indicated are half-lives. This was the state of knowledge at the beginning of the twentieth century.

Source: Joly, J. 1909. *Radioactivity and Geology*. Constable & Co. London.

Holmes–Houtermans model, although Gerling clearly takes prece-
dence (his paper was in Russian). Houtermans was unaware of Holmes's paper until his own paper went to press.

The Gerling–Holmes–Houtermans approach, based on Nier's
isotope data, made the following assumptions (although they were
not explicitly stated):

- The Earth initially had a uniform lead isotopic composition up until the
  time the crust formed.
- After the crust formed, the Pb/U ratio varied slightly from place to place.
- The Pb/U ratios were frozen at different places and did not change except
  due to radioactive decay.
- The lead in each ore mineral did not mix with the lead from other places.

The model corrects the lead isotopes of radiogenic minerals by
subtracting the primeval lead composition using Nier's samples that
showed the least radiogenic lead. Additionally, the model assumes
that these least radiogenic leads represented primeval or initial lead
for the Earth. In a more recent synthesis entitled Plumbotectonics
(plumb, Latin for lead) involving hundreds of lead isotope analyses
from a wide variety of geologic environments, the authors showed
that most lead isotopes can be fairly well modeled based solely on the
three distinct reservoirs with different Pb/U ratios – the lower continental crust, the upper continental crust and the Earth's mantle.[29]

It is possible to calculate three independent isotopic ages from
the three decay schemes noted above – Arthur Holmes used two of
these decay schemes, namely the ratio of the lead-207 decay scheme
and the lead-206 scheme. The advantage of taking this ratio is that
information on the uranium isotope abundances is not required
(see Box 3.1). His results for the age of the Earth were between 2.7
and 3.3 billion years, with different sample combinations giving different ages.[26] It was clear that many of Nier's samples did not fulfill
the strict assumptions required, namely that the Pb/U ratio remain
constant during geologic time. Igneous activity involving melting of
rocks is likely to change this ratio over time, and this is probably why

BOX 3.1 **Age of Meteorites and the Earth**

The law of radioactive decay (Chapter 2) can be solved to give:

$$N_d = N_p(e^{\lambda t} - 1),$$

where N_d is the number of radiogenic daughter atoms, and N_p is the number of parent atoms remaining after time t. We must add the initial number of daughter atoms present N_i to the total number of daughter atoms (N_d):

$$N_d = N_i + N_p(e^{\lambda t} - 1).$$

In the case of the decay scheme $^{235}U \rightarrow ^{207}Pb$, we have the following equation, after normalizing to the stable lead isotope, ^{204}Pb:

$$^{207}Pb/^{204}Pb = \left(^{207}Pb/^{204}Pb\right)_i + \left(^{235}U/^{204}Pb\right)(e^{\lambda t} - 1).$$

An identical equation can also be written for ^{206}Pb decay scheme from ^{238}U. Since it was known in the 1930s that the modern ratio of ^{235}U to ^{238}U was constant at about 1/138, we can eliminate uranium isotopes by taking the ratio of these two equations and putting A_i and B_i for the initial ratios:

$$\left[^{207}Pb/^{204}Pb\right]/\left[^{206}Pb/^{204}Pb\right] = A_i/B_i + 1/138[(e^{\lambda_1 T} - 1)/(e^{\lambda_2 T} - 1)],$$

where λ_1 and λ_2 are the half lives of ^{235}U and ^{238}U respectively and T is the age of the Earth. This equation is important in the history of research on the age of the Earth. Nier was the first to use this equation in his 1939 paper on his studies of lead isotopes in ore minerals; it was also used by Gerling (1942), and by Holmes and Houterman independently in their 1946 papers for their estimates of the age of the Earth. Patterson used it in his 1955 and 1956 papers to date meteorites and the Earth. The popularity of this equation, as noted already, is that the measurement of uranium isotope abundances is not required. The equation is a straight line on a graph where the y-axis is represented by $^{207}Pb/^{204}Pb$, and the x-axis is $^{206}Pb/^{204}Pb$. When time equals zero (T = 0), A_i/B_i represents the x-y coordinates of the initial or primeval lead composition of the Earth. Houtermans used the term "isochron" for the first time to describe points that fall on this line, and today this diagram is called a Pb–Pb isochron (Figure 3.3).

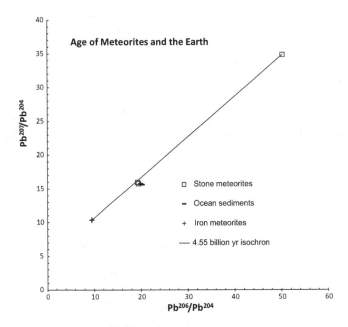

FIGURE 3.3 A Pb–Pb isochron for meteorites and modern ocean sediments.
The slope of the isochron indicates an age of 4.55 billion years, the
currently accepted age of the Earth.
Source: Patterson, C. 1956. Age of meteorites and the Earth. *Geochimica et
Cosmochimica Acta*, v. 10, 230–237. Courtesy of Elsevier.

Holmes' estimate was too young by about one billion years. In addition, Nier's least radiogenic samples likely did not represent the primeval or initial Earth lead composition. The calculations done independently by Cerling and Houtermans (also using Nier's data), although not as extensive as those of Holmes, gave similar results.

The next important attempt to estimate the age of the Earth was made by Claire Paterson and his colleagues for a work published in 1955[30] and 1956.[31] They made two bold assumption: First, that the lead isotopic composition of iron meteorites represented the primeval or initial lead composition of the Earth. They further suggested that the lead isotopic composition of modern ocean floor sediments were average mixtures for the Earth as a whole. They dated five meteorites (three stone meteorites and two iron meteorites) to give an age of

4.55 billion years using the Pb–Pb isochron method (see Box 3.1). They then showed that modern ocean sediments also lie on the same meteorite isochron, implying that meteorites and the Earth were part of the same lead isotopic system and had the same age: namely 4.55 billion years. This is the currently accepted age of the Earth (Figure 3.3).

REFERENCES

1. Kelvin, Lord (Thompson, W.). 1864. On the secular cooling of the Earth. *Trans. Roy. Soc. Edinburgh*, v. 23, 157–169.
2. Lyell, Charles. 1837. *Principles of Geology: being an inquiry how far the former changes of the Earth's surface are referable to causes now in operation.* 1st American edition. Kay & Co., Pittsburgh.
3. Huttton, J. 1788. On the theory of the Earth; or an investigation of the laws observable in the composition, dissolution and restoration of the globe. *Transactions of Royal Society of Edinburgh*, v. 1, 209–304.
4. Kelvin, Lord (Thompson, W.). 1861. On the age of the Sun's heat. *Macmillan's Magazine*, v. 5, 388–393.
5. Buchwald, J. Z. 2008. Kelvin, Lord (Thompson, W.). *Complete Dictionary of Scientific Biography*, v. 13, 374–388. Charles Scribner's Sons, Detroit.
6. Kelvin, Lord (Thompson, W.). 1897. *The Age of the Earth as an Abode Fitted for Life*. 337–356 Annual Report of the Smithsonian Institution.
7. Leibniz, G. W. 1693. *Protogaea*. Translated by C. Cohen and A. Wakefield, 2008. Chicago University Press.
8. Burchfield, J. D. 1975. *Kelvin and the Age of the Earth.* Science History Publications, New York.
9. Aldrich, M. L. 2008. King, Clarence. *Complete Dictionary of Scientific Biography*, v. 7, 370–371. Charles Scribner's Sons, Detroit.
10. Barus, C. 1892. The relation of melting point to pressure in case of igneous rock fusion. *American Journal of Science*, v. 43, 253–254.
11. King, C. 1893. The Age of the Earth. *American Journal of Science*, v. 55, 265–270.
12. Becker, G. F. 1910. Halley on the Age of the Earth. *Science*, v. 31, 459–461.
13. Dalrymple, G. B. 1991. *The Age of the Earth*. Stanford University Press, Stanford.
14. Eyles, E. 2008. John Joly. *Complete Complete Dictionary of Scientific Biography*, v. 7, 160–161. Charles Scribner's Sons, Detroit.

15. Joly, J. 1909. *Radioactivity and Geology.* Constable & Co, London.

16. Joly, J. 1899. A geological estimate of the age of the Earth. *Annual Report of the Smithsonian Institute,* 247–288.

17. Joly, J. 1923. Age of the Earth. *Scientific Monthly,* v. 16, 205–216.

18. Drake, C. L., Imbrie, J., Knauss, J. A. and Turkian, K. K., 1978. *Oceanography.* Holt, Rinehart & Winston, New York.

19. Holmes, A. 1913. *The Age of the Earth.* Harper & Brothers, London.

20. Edmonds, J. M. 2008. William Sollas. *Complete Complete Dictionary of Scientific Biography,* v. 12, 519 – 520. Charles Scribner's Sons, Detroit.

21. Sollas, W. J. 1909. Anniversary address of the President. *Proceedings of the Geological Society of London,* v. 65, 71–123.

22. Nier, A. O. 1938. Variations in the relative abundance of the isotopes of common lead from various sources. *Journal of American Chemical Society,* v. 60, 1571–1576.

23. Nier, A. O. 1981. Some reminiscences of isotopes, geochronology, and mass spectrometry. *Annual Review of Earth Planetary Sciences,* v. 9, 1–17.

24. Nier, A. O. 1939. The Isotopic constitution of radiogenic leads and the measurement of geologic time. *Part II. Physical Review,* v. 55, 153–163.

25. Gerling, E. K. 1942. Age of the Earth according to radioactivity data. *Doklady akademii nauk SSSR, Comptes rendus de l'Académie des sciences de IURSS,* v. 34, 259–261.

26. Holmes, A. 1946. An estimate of the Age of the Earth. *Nature,* v. 157, 680–684.

27. Holmes, A. 1947. A revised estimate for the Age of the Earth. *Nature,* v. 159, 127–128.

28. Houtermans, F. G. 1946 (1975). The isotopic abundances in natural lead and the age of uranium. In *Benchmark Papers in Geology, Geochronology: Radiometric Dating of Rocks and Minerals,* Harper, C. T. (ed.). Dowden, Hutchinson & Ross, Pennsylvania. Translated from the German.

29. Doe, B. R. and Zartman, R. E. 1979. Plumbotectonics, the Phanerozoic. In *Geochemistry of Hydrothermal Ore Deposits,* Barnes, H. L (ed.), Wiley, New York.

30. Paterson, C., Tilton, G. and Inghram, M. 1955. Age of the Earth. *Science,* v. 121, 69–75.

31. Patterson, C. 1956. Age of meteorites and the Earth. *Geochimica et Cosmochimica Acta,* v. 10, 230–237.

4 The Origin of Igneous Rocks

*In regard to basalt and other igneous rocks, Werner's theory was original,
but it was also extremely erroneous.*

– Charles Lyell, 1830[3]

THE NEPTUNIST–PLUTONIST CONTROVERSY

In the previous chapter we saw how geologists made substantial
progress with fossiliferous sedimentary rocks in constructing the
Phanerozoic geologic timescale. But the study of igneous rocks, in
contrast, was substantially retarded in the late eighteenth and early
nineteenth centuries, thanks mainly to the Wernian or Neptunist
school of thought, which believed that basalt and granite were of
aqueous origin.[1]

Educated men of science (and it was almost entirely men at
that time; William Buckland's wife however did contribute substantially to his research and publications) were aware of the AD 79 eruption of Mt. Vesuvius that embalmed the city of Pompeii, Italy, in
volcanic ash, and the more recent eighteenth- and nineteenth-century
eruptions of Mt. Etna in Sicily. For the gentry and men of science
alike, these volcanic sites were usually on the itinerary of their "grand
tour," so that many of these savants were familiar with volcanic
igneous activity firsthand. Sir James Hall, a friend of James Hutton,
for example, took volcanic samples at Vesuvius for use in his melting
experiments at Edinburgh.[2] Lyell himself visited Mt. Etna in 1819 and
described trees that were burned by the 1810 eruption.[3] He also
studied the surrounding marine strata of Sicily, which contained
mollusks similar to those still living in the Mediterranean. These
strata must have been raised several thousand feet above sea level
relatively recently. He speculated that the strata were raised by intrusion of deeper level plutonic rocks that underlie Mt. Etna that were

also the source of the volcanic activity, an idea which Hutton would have agreed with (Chapter 1).

The Neptunists, in contrast, maintained that the source of subterranean heat for volcanic activity was due to burning coal beds at depth, even though most volcanic activity was not associated with coal.[1] Uplifted fossiliferous strata were due to precipitation from a global primordial ocean with an uneven bottom. Where this global ocean came from or where it went was not addressed, but the biblical story of Noah's flood was not explicitly invoked, making the role of religion on this school of thought difficult to evaluate. Many of the clergy nevertheless favored Neptunism compared to Hutton's volcanism. Their global ocean had the miraculous ability to precipitate limestone, granite, and basalt at will, thereby solving all petrological problems.[4]

Hutton, as the leading advocate of the Plutonist school was familiar with the intrusive basalts of Salisbury Craigs near his home in Edinburgh. He saw igneous activity as a force for renewal of the land countering the forces of erosion.[5] Lava flows were clearly due to surface volcanic eruptions, but basaltic sills were commonly interleaved with horizontal sedimentary strata. Hutton correctly recognized them as basaltic lavas intruding horizontally at depth into submarine sediments, but Werner, the German professor, and his followers saw these basalts as sediments. For the Neptunists, lava flows and basalt were different rock types: the former igneous, the latter sedimentary. The other major fiction the followers of the Saxon professor maintained was that granite was also a sediment precipitated from the universal ocean. It's unfossiferous nature being due to its antiquity as the oldest rock type. But Hutton had shown that granite veins intruded into schist in Scotland in 1794, and were therefore younger than the schists (Chapter 1).

An important study of volcanic rocks was made by the French scientist Nicolas Desmarest (1725–1815), who produced detailed maps showing lava flows emanating from extinct volcanic centers at Auvergne near the town of Clermont-Ferrand in south-central France.[6]

This was the first discovery of extinct volcanic activity in France, connecting lava flows with extinct volcanic cones and craters. Moreover, the volcanics were underlain by granite, making coal beds an unlikely source of heat that was the widely held belief of the Neptunists. Desmarest, who lived to be 90 years old, when challenged by Neptunists on his work was not tempted into the argument; he simply replied, "Go and see."[3]

Desmarest was born in northern France, and after he got a college education he went to Paris around 1746 to pursue a scientific career.[7] In 1751 he entered an essay competition sponsored by the French Academy of Sciences. The topic of the competition was whether there had been a land bridge between England and France. He won the prize, and afterward concentrated on physical geography and rocks while supporting himself with government jobs related to the regulation of industry, a role that allowed him to travel widely throughout France. As part of his official duties, he visited Auvergne in 1763 and noticed that the surrounding rocks were prismatic volcanic columns, similar to those of the Giant's Causeway in County Antrim, Ireland, which was one of the most famous geological localities in Europe at that time.

A second locality close to the Giant's Causeway at Portrush also became famous in the Plutonist versus Neptunist debate. At Portrush, a basaltic sill intrudes lower Jurassic shale and contains ammonite fossils, an extinct type of coiled mollusk. Close to the contact, the shale is baked and looks like basalt. One of the Irish Neptunists at the time was the Reverend William Richardson, who sampled fossiliferous shale near the contact and called it "fossiliferous basalt" as proof of the sedimentary origin of basalt. He circulated samples to many geologists, including Playfair and Sir James Hall in Edinburgh. Playfair later visited the site in 1802 and recognized the fossiliferous rock as indurated shale, and not basalt.[8]

Desmarest published his first detailed map of part of the Auvergne volcanics in 1771 together with a treatise entitled "Mémoire sur l'origine et la nature du basalte a grandes colonnes polygons"

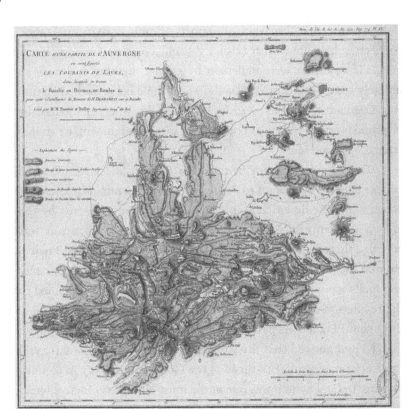

FIGURE 4.1 Map of the Auvergne volcanic region in south-central France showing lava flows of different ages and the locations of prismatic basalts. Accompanied by a manuscript by Nicolas Desmarest.
Source: Desmarest, N. 1771. Mémoire sur l'origine & la nature du basalte à grandes colonnes polygones, déterminées par l'histoire naturelle de cette pierre, observée en Auvergne. Mémoires de l'Académie Royale des Sciences, 705–775.

(An account on the origin and nature of large polygonal columnar basalts).[6] The map shows the locations of columnar basalts, and individual lava flows are distinguished as modern or ancient with arrows indicating the direction of flow (Figure 4.1). The lavas flow in a radial pattern away from extinct volcanic cones and craters. Desmarest himself never became part of the Plutonist school, not believing in a deep source for volcanic rocks.[7] He was accepted into the French Academy of Sciences in 1771 on account of his work at Auvergne. Today we know the Auvergne volcanics are young, Pliocene to Recent in age (from 4 million to 5 million years old) to as young as several hundred thousand years old. Desmarest's maps correctly captured the distinction between young and old volcanic activity.

Jean-Franscois d'Aubuisson de Voisins (1769–1841), a distinguished student of Werner, began studying the basalts of Saxony in the Hartz mountains of northern Germany, the heartland of the Wernian school. He was aware of Desmarest's work, but had not seen those rocks himself and wanted to evaluate the origin of the Saxony basalts. D'Aubuisson published his "Memoire sur les basalts de la Saxe" (An account of the basalts of Saxony) in 1803, which was translated into English in 1814 and re-printed by Cambridge University Press in 2013.[4] In it, he firmly concluded that the Saxony basalts were a chemical precipitate in line with his mentor's doctrine. He also states in his memoir that he had never seen recent volcanic lava flows, such as those of Auvergne or those of Mt. Etna. After accepting his manuscript for publication in 1803, the reviewers recommended that he should go and see the volcanics of Auvergne for himself, which he did in the summer of 1804.[1] Once there, he was convinced of the volcanic origin of the rocks of Auvergne. In the English version of his memoir (1814), the translator nevertheless notes d'Aubuisson still maintained the Saxony basalts were of sedimentary origin. (Incidentally, he was aware of Richardson's mistake regarding the Portrush fossiliferous shale mentioned earlier.)

It is useful to summarize some of d'Aubuisson's arguments for a non-volcanic origin of the Saxony basalts, as it gives some insight into the Neptunian school of thought. The basalts he examined all capped the surrounding mountaintops in the region at about the same level, and he reasoned that they had once been part of a much larger sheet, too large to be ejected from a single volcano – and multiple volcanic sources were not to be found. Moreover, he noted that the primitive rocks (or basement rocks in modern parlance) of granite and gneiss were penetrated by mine shafts in the search of ore bodies, some of which he explored, but nowhere could vertical volcanic feeder pipes be seen. He further noted the absence of cinders or scoria typical of modern lava flows. He noted the basalt should be observed to flow downhill and not maintain a constant elevation. All of these comparisons with modern lava flows were not relevant because the Saxony basalts are shallow horizontal intrusions into sedimentary strata (sills), similar to those at Edinburgh. Much of the Plutonist-Neptunist controversy can be ascribed to comparing surface lava flows to basaltic intrusions or sills.

According to the Neptunists, the apparent gradational nature of the basalt with surrounding sediments was a sure indication of their sedimentary origin. D'Aubuisson also notes that he and others see no heating effects of the basalt with the surrounding sediments, including coal. He compared the chemistry of basalt from Mt. Etna with his basalts and also with graywacke, a type of muddy sandstone. The basalts and graywacke contained 5 percent water, whereas the Mt. Etna volcanics contained none, and this, he argued, favored an aqueous origin for basalt. He cannot have been expected to know at this time that weathering introduces hydrous minerals to basalts and other rocks, whereas the Mt. Vesuvius lavas were fresh. He does, however, ignore the fact that the Al, Na, Ca, and Fe contents of the Mt. Etna lava were very similar to his Saxon basalt. He also disregarded the results of Hall's experiments, which showed that slow cooling of a molten glass produced basaltic textures. He concluded that precipitation from the primordial ocean can have been the only possible source of basalt.

When asked what type of solvent was the global "menstruum" and what caused the precipitations d'Aubuisson, replied, "I do not interfere with remote causes but confine myself to the facts."[4] But, he eventually concedes in his response, "It must be analogous to our present seas."[4] Although the Neptunists knew the chemical composition of the rock-forming minerals and also that of seawater, they never used chemical reactions to try explain their precipitation theory. Werner's school at Freiberg, Germany, mainly gave instruction on mineralogy, mining, and "geognosy" (their term for geology), although Werner's biography says he kept abreast of chemical developments.[7]

There are several reasons why Neptunists and the Wernian school of geognosy dominated over the Plutonist or Plutonist school for so long in the late eighteenth and early nineteenth centuries. One is that Werner had many distinguished and ardent students who went forth and preached the Neptunist faith abroad. Robert Jameson (1774–1854) was one of Werner's students who became professor at Edinburgh in 1804 and there founded the Wernerian Natural History Society.[7] In his memoir, D'Aubuisson cites how he visited Scotland on a field trip with Jameson to examine coastal outcrops. He was delighted to see over forty eager geology students on this trip taking notes on the apparent gradational relations between basalt and sediments, proving their sedimentary nature.

In contrast, Hutton, the leading advocate of the Plutonist school, did not hold an academic teaching position; besides, as Steven J. Gould noted in his book *Time's Arrow*, he was a "lousy" writer. His friend and supporter John Playfair was a professor of mathematics, not of geology, at Edinburgh. Alexander von Humboldt (1769–1859), another student of Werner, was one of the most traveled and famous scientists of his day. Werner's narrow-minded and blinkered view of the geological world, which was focused solely on the geology of Saxony, an area he never left, went forth and multiplied with the help of his students, until it was eventually disowned by geologists in Europe and in America.[3] Archibald Geikie, in his book *The Founders of Geology* (1905), referred to Werner as "disastrous to the higher interests of geology."[1]

THE PETROGRAPHIC MICROSCOPE

The use of the microscope, invented earlier in the seventeenth century, was slow to gain acceptance among the geological community until the mid-nineteenth century. But once it was adopted, microscopic studies confirmed the fallacy of Neptunism once and for all. William Nicol (1768–1851) was an instructor in the physical sciences (specifically natural philosophy) at the University of Edinburgh, and his interests were geological in nature. He made two important technical inventions that were to prove crucial in the development of the petrological microscope. He developed the technique of making thin sections in 1815, which was not published until much later, whereby rock slices were thinned and polished so that they could be examined under transmitted light. He also invented the Nicol prism, which allowed rocks in thin sections to be viewed in polarized light under the microscope. Together, these developments proved very powerful in allowing minerals and textures to be identified under the microscope. Geologists were slow to make use of these new research opportunities, and Geike noted in this regard that "geologists are as about as difficult to move as their own erratic boulders."[1] Forty years later, in 1858, Henry Sorby used Nicol's inventions and published an important paper on microscopic fluid inclusions in rocks, starting an important subdiscipline in petrology that is still very active today, namely fluid inclusion research.[9] Sorby, in his paper, encouraged skeptical geologists to take up microscopic petrology, some of whom noted "it is impossible to look at a mountain through a microscope."[9] A German geology student and friend of Sorby, Ferdinand Zirkel (1838–1912), needed no such encouragement; he studied the volcanic rocks of Iceland using the petrographic microscope and obtained his Ph.D. in 1861 from the University of Bonn for this work. He also produced the first authoritative treatise, shortly thereafter, on microscopic petrology, establishing it as an important discipline within geology. When geologists did finally take up microscopy, they saw that basaltic sills and modern lavas had the same textures and minerals, and probably therefore had the same igneous origin (Figure 4.2). Had geologists taken up the petrologic microscope earlier, Neptunism would not have lasted so long.

The Neptunist–Plutonist controversy was mainly about one rock type: basalt. The wider Neptunist–Plutonist controversy included the origin of granite. The Neptunists not only thought granite was the oldest rock type, naming it part of their Primitive series along with gneiss and metamorphic rocks, but considered it as sedimentary and attributed the absence of fossils in granite to its ancient origin before

FIGURE 4.2 Volcanic rocks under the petrographic microscope under different magnifications. 1) Lava from Mt. Etna showing augite (dark), feldspar (colorless), and iron oxides. 2) Lava from Mt. Vesuvius's AD 79 eruption on Herculaneum showing dark, crystal augite and feldspar colorless. 3) Basalt from Tahitii showing crystals of olivine, pyroxene, and feldspar. 4) Obsidian from Arran Island, Scotland, showing fine crystals and pyroxene.

Source: Forbes, D. 1867. The microscope in geology. Popular Science Review, v. 6, 355–368.

life began.[1] In 1794 Hutton presented a paper to the Royal Society of Edinburgh entitled "Observations on Granite" in which he described visiting a locality in the Grampian mountains of Scotland, along the Glen Tilt streambed, where he saw, with some friends, the intrusive contact between Silurian-aged granite and the older surrounding schist.[11] He brought back samples including large stream boulders with red veins of granite running through dark-colored schist. This observation proved that granite was younger than the surrounding rocks, and also that it had an igneous intrusive origin.

The Neptunist school, having declined in influence, allowed geologists to pursue more important issues such as explaining the wide diversity of igneous rocks they encountered as igneous provinces around the world became better known geologically. Much of the early work on igneous rocks involved description of different rock types with a tendency to label slightly different rocks with a new name, often after a specific locality, leading to a very burdensome nomenclature. The American geologist Reginald Daly noted in his book *Igneous Rocks and Their Origin* (1914), that there existed 770 rock names.[12] In an appendix to *Petrology of Igneous Rocks* (1926), the petrologists Hatch and Wells listed 660 rock names and noted with British understatement, "The bulk of them are unnecessary."[13]

DIFFERENTIATION MECHANISMS AND ROCK CLASSIFICATION

Based on examination of rocks under the microscopic in the 1870s and the increasing use of chemical analysis at this time, classification schemes emerged to group rocks into igneous suites, variously called families or clans, the members of which would prove, hopefully, to be related to each other (e.g., basalt→ andesite→ rhyolite) and thereby render the profusion of so many rock names unnecessary. The English petrologist Alfred Harker, in his book A *Natural History of Igneous Rocks* (1909), proposed a useful way of portraying chemical data, in which weight percent silica was plotted on the horizontal axis and the other oxides were plotted on the vertical axis (Figure 4.3). The resulting curves either tended to be smooth (or were smoothed by Harker), becoming nearly linear and could be used to address a variety of scientific hypotheses as to how a series of related rocks could be produced, whether by mixing of two magmas of different composition

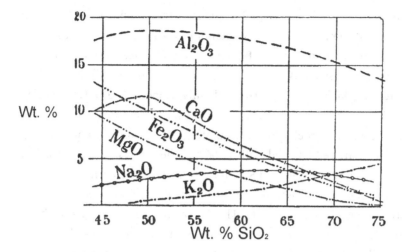

FIGURE 4.3 Harker variation diagram for volcanic rocks of Lassen Peak, California. This presentation of rock chemistry is useful in examining rocks belonging to the same magmatic series.
Source: Harker, Alfred. 1909. *The Natural History of Igneous Rocks.* Cambridge University Press.

or assimilation of country rock sediments or by crystal fractionation. So, for example, a sialic magma with 75 percent silica and 0 percent iron and magnesium in Figure 4.3 could be mixed with a mafic magma with 50 percent silica in various proportions to produce the series of rock compositions observed. The experimental petrologist N. L. Bowen argued in his book *The Evolution of Igneous Rocks* (1928) that Harker variation diagrams, as they came to be known, strongly indicate the role of fractional crystallization in producing the variations observed, especially in volcanic rocks since they most likely reflected a magmatic liquid line of descent.[15] But other interpretations are possible.

With such a wide variety of igneous rock types with all gradations between them, the classification of igneous rocks proved to be exceptionally challenging. The petrographic microscope, however, advanced rock classification greatly, particularly in the case of volcanic rocks because of their fine grain size. Geologists in the field needed a classification of plutonic rocks based on the modal mineralogy, but some researchers wanted a quantitative classification based on chemical composition. The mnemonic term *mafic* refers to rocks which are rich in Mg and Fe, and *sialic*, also mnemonic, refers to rocks rich in Si and Al. The old terms acid and basic, still used today, were introduced by the German chemist Robert Bunsen (1811–1899) for rocks rich and poor, respectively, in silica.[7] One such classification divides igneous rocks into ultramafic ($<45\%$ SiO_2), mafic ($<52\%$), intermediate ($<66\%$) and sialic ($>66\%$). The German petrologist Karl Rosenbusch (1836–1914) of Strasburg University recommended a classification of rocks based on three criteria, namely mineralogy, chemistry, and geological context.[7] The South African petrologist S. Shand noted that certain minerals are incompatible with one another.[16] For example, olivine and the feldspathoids do not co-occur with quartz because they react with quartz to form pyroxene and alkali feldspar, respectively. In the case of olivine the reaction is:

$$Mg_2SiO_4 + SiO_2 \rightarrow Mg_2Si_2O_6$$

$$Olivine + quartz \rightarrow pyroxene$$

For rocks with free quartz, Shand termed them *oversaturated*; those with no free quartz, he called *saturated*; and those with olivine and feldspathoids, he called *undersaturated*. Johannsen's classification published in 1917 further subdivides igneous rocks based on the ratio of alkali feldspar to calcic feldspar, an idea originally suggested by German petrologists that reflects the recognition early on of the existence of alkaline and calc-alkaline igneous rock suites as distinct families.[17]

Eventually, after much discussion and debate (as recently as 1976), the International Union of Geological Sciences agreed on a modal classification of coarse-grained igneous rocks that is based partly on earlier work by German petrologists and Johannsen's classification scheme (Figure 4.4). Shand's ideas also likely played a role in the design of this classification because the top of this diagram is reserved for quartz oversaturated rocks and the bottom for quartz undersaturated rocks. A separate triangle was required for ultramafic rocks.[18]

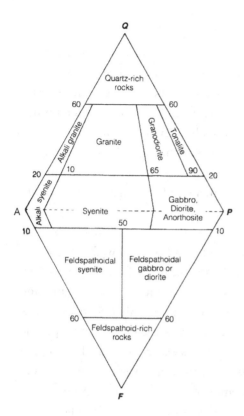

FIGURE 4.4 Classification of plutonic rocks. Q: quartz; A: alkali feldspar; P: plagioclase; F: feldspathoids. A separate triangular diagram is used for ultramafic rocks (not shown). *Source:* Streckeisen, A. 1976. To each plutonic rock its proper name. *Earth Science Reviews*, v. 12, 1–33. Courtesy of Elsevier.

An influential American classification, introduced in 1903, based on chemical composition should also be mentioned, namely the CIPW norm, named for the authors' last name initial letter (Cross, Iddings, Pirsson and Washington).[19] In this classification, the chemical components of the rock chemistry are apportioned to different minerals in a specific order resulting in a "normative" mineralogy, which may differ from the actual modal mineralogy. This allows, for example, the comparison of normative mineralogy of a gabbro and a fine-grained basalt which can be quite useful. The CIPW norm calculation was very cumbersome before calculators became common but is still used today, mainly using computer routines, over a hundred years since its original proposal.

EXPERIMENTAL PETROLOGY

The main goal of igneous petrology is to explain the mechanisms whereby the wide diversity of igneous rocks could be produced from one or two parental magma types. It was widely thought in the early twentieth century that the sialic (granitic) crust was underlain by a mafic (basaltic) substratum providing two end-member magma types.[12] Differentiation mechanisms included assimilation of country rocks by an intruding magma[13], magma immiscibility, magma convection, magma mixing[14], fusion of sediments, and crystal fractionation.'[15]

The idea of crystal fractionation was observed early on by visitors to Hawaii, where the bottom of lava flows showed an agglomeration of feldspar crystals while the surface of the flow showed no such crystals. The mechanism involves separation of liquid magma from crystals often by gravitational settling. In the United States, the chief proponent of fractional crystallization was the experimental petrologist Norman L. Bowen (1887–1956). His experiments at the Geophysical Laboratory of the Carnegie Institution, Washington DC, were crucial to showing the importance of this mechanism. A quote from his *Evolution of Igneous Rocks* (1928) is revealing: "Of all the hypotheses of differentiation of magmas none except the hypothesis of fractional crystallization can be checked against observation in any detail. It is therefore the only one that can be regarded as having any sound scientific basis."[15] Today, this can be regarded as a gross overstatement, but at the time his opponents were proposing various hypotheses based largely on field studies that were difficult to test (and therefore not scientific), especially with regard to the origin of granite.

Bowen's experiments showed that there was a definite sequence of crystallization in a cooling magma (now called Bowen's reaction series) – a continuous sequence involving the plagioclase feldspars, beginning with calcic feldspar crystallizing first, which is also more dense, followed by more sodic and less dense plagioclase. There also existed, simultaneously, a discontinuous sequence of crystallization starting with olivine, followed by pyroxene, amphibole, biotite, and finally alkali feldspar, quartz, and muscovite.[15] Reactions similar to that shown above help explain this discontinuous series. The early high-temperature minerals would crystallize and therefore could produce ultramafic rock types (such as dunite and pyroxenite), and the last-formed minerals could form granite (alkali feldspar and quartz). Bowen, following Reginald Daly before him, his onetime teacher, assumed that the parental magma was basaltic in composition. The study of layered ultramafic intrusions such as the Skaergaard of Greenland appeared to confirm Bowen's experimental results.[20] These ultramafic intrusions contained layered accumulations of ultramafic minerals at their base and increasingly more sialic (granitic) rocks toward the top, suggesting crystal fractionation involving gravity occurred based on mineral densities.

In 1937 Bowen summarized much of his laboratory's experimental work over the previous few decades and showed that cooling and crystallization of basaltic magma produced end products that were progressively enriched in alkali feldspars and feldspathoids.[21] On a triangular diagram involving these minerals and quartz, experiments showed an elongated temperature valley in the center of the plot (see Figure 4.4). Bowen showed that the world's average alkaline volcanic rocks, and specifically those of the East African rift, plotted within the same temperature trough, indicating that crystal-liquid fractionation of a basaltic magma was the likely origin of alkaline volcanic rocks worldwide. Bowen referred to his triangular system as "petrogeny's residual system"[21] because the liquid line of descent of cooling magmatic liquids always ended up in a temperature valley starting from a wide variety of initial compositions.

ORIGIN OF GRANITE

The origin of granite was one of the main preoccupations of petrologists during the 1940s and 1950s, and the arguments were pitted mainly between European field geologists and American petrologists.[22] The Europeans favored metamorphism of sediment as an

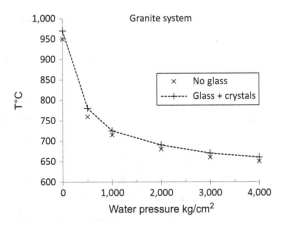

FIGURE 4.5 Plot of temperature versus water pressure for the beginning of melting in the granite system. The temperature of melting decreases, as indicated by the presence of glass, with increasing water vapour pressure. *Source:* Tuttle, O. F. and Bowen, N. L. 1958. Origin of granite in the light of experimental studies in the system $NaAlSi_3O_8$-$KAlSi_3O_8$-SiO_2-H_2O. *Geological Society. America Memoir, v.* 74, 153.

origin of granite (transformationists) whereas Americans favored an igneous origin based on magmatic differentiation (magmatists). Many European granites contain inclusions (or xenoliths) of sedimentary material, evidence favoring a sedimentary origin.

The importance of Bowen's residual system can be further appreciated when it is applied to the granite problem. The origin of granite can be considered when only the upper half of his triangular system is considered, namely quartz at the top and the alkali feldspars (albite and orthoclase) along the base of the triangle (Figure 4.4). Further experiments by Tuttle and Bowen (1958) in this system examined the effect of water pressure on the thermal valley. This research first required development of a new hydrothermal pressure apparatus. The experiments then took an additional five years, but the results were startling (Figure 4.5).[23]

In the granite system, the temperature valley dropped from 970°C under dry conditions to 660°C at high water pressure. Water under high pressure had the effect of a flux, lowering the melting

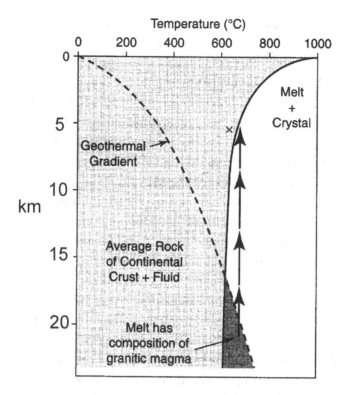

FIGURE 4.6 Diagram showing intersection of the continental geothermal gradient with the melting curve for wet granite. Granite magmas can form in the crust between a depth of 10 kilometers and 25 kilometers, depending on the geothermal gradient. This allows the melting of crustal sediments as a major source for granites.

Source: Skinner, B. J. and Porter, S. C. 1989. *Dynamic Earth*. Wiley, New Jersey. Courtesy of Wiley.

temperature very substantially. Furthermore, this temperature valley corresponded precisely to the modal composition of alkali feldspar granites suggesting a magmatic origin for such granite. An alternative origin for granite, however, also immediately suggested itself.

As Tuttle himself stated, "[T]he melting of sediments deep in the crust also now became a primary source of granitic magma."[23] (See Figure 4.6.) In other words, the melting of the thick geosynclinal sediments of future orogenic belts (10–20 kilometers deep) could be the source of granite batholiths, as European geologists had long maintained. Some granites can form by crystal fractionation and others by partial melting of sediments, confirming the English geologist Herbert Read's earlier (1948) statement: "[T]here are granites and granites."[22] Bowen died in 1956 before the publication of the Tuttle and Bowen memoir in 1958. His 1928 book nevertheless remains the most important book on igneous petrology of the twentieth century.

In a fiftieth anniversary volume (1979) dedicated to Bowen, one of the contributors compared the granitic rocks of the Skaergaard ultramafic intrusion (formed by crystal fractionation) to granitic rocks of the Sierra Nevada and found little correspondence between the two.[24] An important question raised was whether or not fractional crystallization could produce granitic mountain ranges such as the Sierra Nevada or those of the Andes in Peru, which are mainly granodiorite and tonalite in composition rather than strictly granite (see Figure 4.4). If these mountain ranges were formed by fractional crystallization, they should be underlain by very large mafic magma chambers. Based on geophysical evidence, mainly seismic and gravity, the absence of basic magmas beneath these granitic mountain ranges indicates that the answer is no (unless these mafic magmas have since disappeared by some hypothetical mechanism). An explanation of these orogenic-scale granitic belts would have to await the discovery of plate tectonics (Chapter 7). The history of igneous petrology has been ably treated by Davis A. Young in *Mind over Magma*.[25]

In summary, there were three main reasons the Neptunist school lasted so long: Werner had many ardent students who promoted his ideas widely; Hutton, in contrast, was not a teacher with an academic position, and he was also a poor writer; lastly, geologists were slow to take up the petrographic microscope, which would have ended Neptunism sooner. The petrographic microscope also helped in developing a classification of igneous rocks, but it was a long time before international agreement was reached on such a classification. The Harker variation diagram played an important role in testing hypotheses of magma differentiation. The experimental petrologist

Bowen championed crystal-liquid fractionation throughout his life as the most important differentiation mechanism. The Tuttle and Bowen experiments on granite showed the importance of water in lowering magma melting temperatures and allowed granites to be produced in the continental crust by melting of sediments, as European geologists had long maintained based on field evidence. The origin of granite is discussed further in Chapters 7 and 8.

REFERENCES

1. Geikie, A. 1905. *The Founders of Geology*, Macmillan, 2nd ed., London.
2. Hall, J. 1798. Experiments on whinstone and lava. *Transactions Royal Society Edinburgh*, v. 5, 44–75.
3. Lyell, C. 1880–1883 (1997). *Principles of Geology* (3 vols.). Penguin Classics, London.
4. D'Aubuisson, J.-F. 1814. *An Account of the Basalts of Saxony: With Observations on the Origin of Basalt in General*. Trans., P. Neill, 2013. Cambridge University Press, Cambridge, UK.
5. Hutton, J. 1788. Theory of the Earth; or an investigation of laws observable in the composition, dissolution and restoration of land upon the globe. *Transactions Royal Society Edinburgh*, v. 1, 209–305.
6. Desmarest, N. 1771. Mémoire sur l'origine & la nature du basalte à grandes colonnes polygones, déterminées par l'histoire naturelle de cette pierre, observée en Auvergne. *Mémoires de l'Académie Royale des Sciences*, 705–775.
7. Taylor, K. L. 2008. Nicolas Desmerest. *Complete Dictionary of Scientific Biography*. Charles Scribner's Sons, Detroit.
8. Davies, G. L. H. 1981. The History of Irish Geology. In *A Geology of Ireland*, C. H. Holland (ed.). Scottish Academic Press, Edinburgh, 303 – 315.
9. Sorby, H. C. 1858. On the microscopic structure of crystals. *Quarterly Journal Geological Society*, v.14, 453.
10. Forbes, D. 1867. The microscope in geology. *Popular Science Review*, v. 6, 355–368.
11. Hutton, J. 1794. Observations on granite. *Transactions Royal Society Edinburgh*, v. 5, 77–81.
12. Daly, R. 1914. *Igneous Rocks and Their Origin*. McGraw-Hill, New York.
13. Hatch, F. H. and Wells, A. K. 1926. *The Petrology of the Igneous Rocks*. Allen & Unwin, London.

14. Harker, Alfred. 1909. *A Natural History of Igneous Rocks.* Cambridge University Press, Cambridge.

15. Bowen, N. L. 1928. *The Evolution of Igneous Rocks.* Princeton Press, New Jersey.

16. Shand, S. J. 1913. On saturated and undersaturated rocks. *Geological Magazine,* 10, 508–513.

17. Johannsen, A. 1917. Suggestions for a quantitative mineralogical classification of igneous rocks. *Geology,* v. 25, 63–97.

18. Streckeisen, A. 1974. Classification and nomenclature of plutonic rocks. *Geologische Rundschau,* v. 63, 773–786.

19. Cross, M., Iddings, J. P., Pirsson, L. V., and Washington, H. S. 1903. *Quantitative Classification of Igneous Rocks.* University of Chicago Press.

20. Wager, L. R. and Deer, W. A. 1939. Geological investigations in East Greenland. Part III. The Petrology of the Skaergaard intrusion, Kangerdlugssuag, East Greenland. *Meddelelser om Grønland,* v. 105, 209.

21. Bowen, N. L. 1937. Recent high-temperature research on silicates and its significance in igneous geology. *American Journal Science.,* v. 33, 1–21.

22. Read, H. H. 1948. Granites and granites. In *Origin of Granites* (Gilluly, J. C., ed.). Geological Society America Memoir, v. 28, 1–19.

23. Tuttle, O. F. and Bowen, N. L. 1958. Origin of granite in the light of experimental studies in the system $NaAlSi_3O_8$-$KAlSi_3O_8$-SiO_2-H_2O. *Geological Society America Memoir,* v. 74, 153.

24. Yoder, H. S. (ed.). 1979. *The Evolution of Igneous Rocks: Fiftieth Anniversary Perspectives.* Princeton University Press, New Jersey.

25. Young, D. A. 2003. *Mind over Magma.* Princeton University Press, New Jersey.

5 Tectonics in Crisis

No one would believe me; they would put me into an asylum.

– Arnold Escher von der Linth (1807–1872)[1]

INTRODUCTION

The words tectonics and architecture are derived from the same Greek root, and tectonics is defined as the architecture of the Earth's crust.[1] As outlined in Chapter 2, nineteenth-century geologists achieved remarkable success in constructing a stratigraphic column and producing a geological timescale, but tectonic geologists, on the other hand, were far less successful. The fault lay not this time with the Neptunists, but rather with the large influx of new geologic observations that were not easily explicable. In Europe geologic societies such as the Geological Society of London and Edinburgh were very active, and in North America newly established state geological surveys in the 1830s provided new data on mountain belts. Alpine nappes and large scale thrust faults were difficult to explain, as were trans-Atlantic geologic similarities.

The main theory for the origin of mountains throughout the eighteenth and nineteenth centuries was that of contraction theory, whereby the radius of a cooling Earth decreased with time causing the crust to become crumpled, thereby forming mountain ranges. This theory persisted even after the discovery of radioactivity at the end of the nineteenth century, which suggested that the Earth was not undergoing long-term cooling. In addition, trans-Atlantic geologic correlations between Europe and North America, and also between South America, Africa, and India, were attributed to foundered continents now occupying the intervening ocean basins, even as the idea of isostasy was gaining credibility. Isostasy precluded the foundering of continents. This led to the idea of land bridges as a desperate attempt to explain the observations. With the rejection of continental drift in the early half of the twentieth century, the contraction theory was still the mainstay of tectonics until plate tectonics prevailed in the 1960s. The influential British geophysicist, Sir Harold Jeffreys, supported contraction theory as late as 1976 in the sixth edition of his book *The Earth*. The contraction theory of mountain belts was not only incorrect, it also had little explanatory power. It could not explain, for example, the well-recognized asymmetry of mountain belts or their preferred orientation, but it nevertheless survived well into the twentieth century. Throughout the nineteenth and twentieth centuries, tectonic geologists were in crisis mode. As such, theories focused on mountain-building and the behavior of the continents were both incoherent and untenable as scientific hypotheses.

EARLY MODELS

The seventeenth-century French philosopher René Descartes (1596–1650), in his Principles of Philosophy (1644), presented a theory for the formation of the Earth that was subsequently adopted by polymath savants such as Gottfried Leibniz and also the mathematical physicist Kelvin (William Thompson).[2] Descartes assumed the Earth was similar to the sun, namely that it was initially a star and it gradually cooled from a molten mass and formed concentric layers according to the density of its elements, forming an atmosphere, an ocean, a rigid crust, and denser layers in the interior that were rich in metals. In his vision, the formation of mountains involved fracturing of the crust and crustal fragments falling inward to produce jumbled slaps, forming mountain peaks, and ocean waters then filled in the lower topography (Figure. 5.1). Descartes's crust consisted merely of "clay, sand and mud," whereas Leibniz and Kelvin came closer to the truth by indicating that the crust was a "crystalline mass," referring to igneous rocks.[3] The assumed long-term cooling of the planet was the origin of the contraction theory of mountain belts.

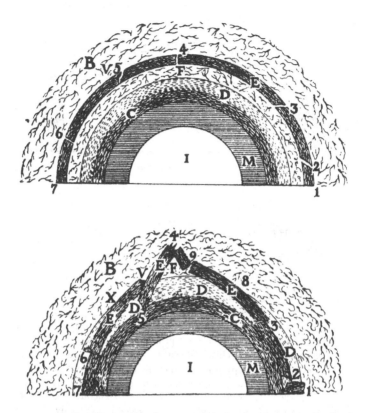

FIGURE 5.1 **Top**: Formation of concentric layers from a molten Earth after it cools. B – atmosphere; D – ocean; E – crust; C, M, and I – deep interior layers. Note that F is also atmosphere, implying the Earth was partially hollow at depth beneath the crust.
Bottom: Cracks develop in the crust (E) and crustal fragments collapse into the void (F). Jumbled crustal fragments produce mountain peaks. The ocean (D) fills in what was previously void (F). The astronomer Edmund Halley also believed that the Earth was partially hollow.
Source: Descartes, René, 1644. *Principles of Philosophy.* Translated by V. Miller and R. Miller, 1983. Riedel Publishing, Dordrecht.

In the mid– to late–nineteenth century, much-needed geologic fieldwork and basic observational data, clearly lacking in earlier theories of the Earth, was beginning to show that several mountain belts had a similar basic structure. One of the earliest papers on this topic was by the Reverend John Mitchell of Cambridge in 1760; although

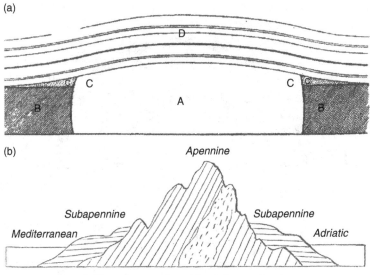

1. *Transverse section of the Italian peninsula*

FIGURE 5.2 (a) One of the earliest general models for a mountain belt in a paper devoted mainly to earthquakes by J. Mitchell (1760). A – mountain core; B – inclined schist and slates; C – overlying strata; D – youngest strata folded by upward motion of the mountain core.
Source: Mitchell, J., 1760. Conjectures concerning the cause and observations upon the phenomena of earthquakes. *Philosophical Transactions*, v. 55, 566–634.
(b) A west-to-east crosssection of the Apennines of Italy after Charles Lyell from his *Principles*. A mountain core of Primitive rock is shown bounded by steeply dipping Secondary strata flanked by unconformably overlying horizontal Tertiary strata.
Source: Lyell, C. 1830–1833 (1997). *Principles of Geology (3 vols.)*. Penguin Books, London.

his paper is mainly about earthquakes, it included a generalized model of mountain belts.[4] The high mountainous ridges were formed by older massive rocks and flanked on either side by steeply dipping younger strata, in turn overlain by nearly horizontal strata that extend away from the mountain ridge for a considerable distance (Figures 5.2(a) and 5.2(b)). As examples of this mountainous ridge structure, the author cited the Andes and the Appalachians. He asks the reader to imagine that the older central core of the mountain ridge is pushed up under the horizontal rocks and is then cut by a horizontal plane of erosion. Mitchell was clearly describing an eroded anticlinal structure overlain by younger undeformed rocks as a general model for mountain belts.

Catherine the Great of Russia (1729–1796) sponsored a very successful scientific expedition to learn about the botany, zoology, and geology of the Russian empire, which expanded greatly under her reign. The survey team included five naturalists, seven astronomers, and seven assistants. The leader of the team was the German naturalist Peter Simon Pallas (1741–1811), who at the time held the chair in natural history at the St. Petersburg Academy of Sciences. The expedition lasted six years (1768–1774) and covered an enormous area starting in St. Petersburg in the west and exploring the Siberian plains east of the Urals and the Altai Mountains in the Far East.[5]

Based on his observations on the Urals, Pallas presented a model of mountain chains similar to that of John Mitchell, but with substantially more geological detail. Granite forms the core of mountains and is covered by unfossiliferous schistose rocks that rest tilted against the granite; these are succeeded by shales and by thick fossiliferous limestones. These fossiliferous tilted strata become less tilted with distance from the mountain core. Like most European geologists at the time, Pallas was a Neptunist and followed the teachings of his fellow countryman, the Saxon professor Abraham Werner, so that all these rocks were thought to have precipitated from a primeval ocean. The granites were unfossiliferous because they were formed before life began, and the fossiliferous rocks were clearly younger. Murchinson on his later trip to Russia relied on Pallas's early observations where Murchinson identified the Permian system near the Ural mountains.[6]

The British tradition of Hutton and Lyell (uniformitarians) did not contribute substantially to the discussion of the origin of mountain belts, but Murchinson and Sedgwick (catastrophists) did publish on the structure and stratigraphy of the Alps, the Carpathians, and the Apennines.[7] Lyell's three volume *Principles* devotes only six pages to mountain belts and presents only a very rudimentary crosssection of the Apennines consistent with Mitchell's and Pallas's model (Figure 5.2(b)).[8] Consistent with his uniformitarian assumption, Lyell argues that the Apennines were raised incrementally by numerous earthquakes associated with volcanism over time. Referring to mountain cores as Primary rocks, the tilted rocks as Secondary, and the horizontal rocks as Tertiary, he noted that the Pyrenees were raised between the age of the Secondary rocks (Cretaceous chalk) and Tertiary rocks. Now that a geologic stratigraphy had been established, mountain belts could be dated stratigraphically. As William Conybeare maintained (Chapter 1), mountain belts such as the Alps did not show evidence for uniformity of nature. This may explain Lyell's reluctance to spend much energy on mountain belt formation.

CONTRACTION THEORY

The early idea that the Earth cooled gradually from a molten ball led to an enduring model for mountain-building that lasted into the twentieth century.[2] The long-term cooling, or the *réfroidissment seculaire* as Fourier called it, caused a contraction of the radius of the globe and therefore corrugation and crumpling of the outer crust as in mountain belts, similar to the skin of a dried apple.[9] This was referred to as the contraction theory, but it was not tested quantitatively until the late nineteenth century and was found wanting. Imminent geologists familiar with mountain belts who supported the contraction theory include the English geologist Henry De la Beche (1796–1855), the American geologist James Dana (1813–1895), and the Alpine geologists Elie de Beaumont (1798–1874) and Eduard Suess (1831–1914).[5]

The theory as stated by geologists was simple, easily understood (the apple analogy), and general enough to be applied to all mountain belts produced by compression, but conversely it had no explanatory power for specific mountain belts. In principle, it was a testable hypothesis, but not enough was known about the Earth's internal thermal history throughout most of the nineteenth century in order to test it; this is perhaps why this theory survived so long. As late as 1878, the geologist Joseph Le Conte was defending the contraction hypothesis against attacks, mainly from the American geologist Clarence Dutton and the English geophysicist Osmond Fisher.[9] Le Conte was aware that compressional mountain belts were asymmetric in structure, but the contraction hypothesis did not predict such asymmetry; he proposed that when asymmetric sedimentary wedges were incorporated into a mountain belt, they would also result in asymmetric mountain structures. We saw in Chapter 3 that Kelvin had shown that the long-term cooling history of the Earth could be calculated, but at that time he was solely interested in a limit on the age of the Earth, not the amount of crustal shortening due to cooling. The calculation of the amount of shortening starts with Kelvin's original calculation, but involves several additional steps making it a substantially more complex mathematical problem. Writing in 1882, in the journal *Nature*, Fisher recounted that the contraction theory occurred to him independently in 1840 and he thought it was such a discovery that "in my youthful joy I vaulted over a gate."[10] Much later, however, when he went to quantify the effect, he found it totally inadequate to account for the observed compression.[11] The radius of the Earth would only shrink 10 kilometers (6 miles), and this would produce a change in elevation above sea level of only 6 meters (20 feet.).

Lyell noted in his *Principles* that the French physicist Laplace had concluded that the Earth had not contracted by cooling; for if it had, the length of the day would have shortened (just as an ice skater spins faster when their arms are lowered due to conservation of angular momentum), but observations since ancient Greek times had showed that this was not so. For Lyell this long-term cooling was inconsistent with his and Hutton's uniformitarian assumption because geologic processes would have been more intense in the past. Kelvin's objection to the uniformitarian assumption was discussed in Chapter 3.

Captain C. E. Dutton (1841–1912) of the United States Geologic Survey observed that a liquid substratum to the crust was required for the contraction theory to work, so that slippage could occur between the deeper levels in the Earth and the contracting overlying brittle crust. In that case, the shrinkage of the crust would not be like an apple, but rather randomly distributed in every direction, which is inconsistent with the observed distribution of mountain chains today, specifically the north-south chains from the Andes to Alaska.[12] Dutton, agreeing with Fisher's's calculations, wrote, "[T]his hypothesis is nothing but a delusion and a snare, and the quicker it is thrown aside and abandoned the better it will be for geological science."[10]

The contraction theory eventually went out of favor not only because of Dutton's criticisms and Fisher's quantitative work, but also because of the very large thrust translations that were being discovered in several mountain belts. In addition, the discovery of radioactivity (1889; Chapter 2) called into question the long-term cooling and contraction of the Earth. The Irish geologist-physicist John Joly (who we briefly met in Chapter 2) was by 1911 considering the role of heating, rather than cooling, in mountain building events.[13] The English geologist, Arthur Holmes, initially supported contraction theory, but by 1925 he said it was inadequate to explain volcanic activity especially in extensional environments.[14] The renowned British geophysicist Sir Harold Jeffreys (1891–1989), however, revived the contraction theory by including heating from radioactivity and concluded in the first (1924) edition of his book *The Earth* that, despite radioactivity, the Earth was still cooling and that the contraction was more than adequate to explain mountain belt shortening[15] (see also Box. 6.1).

GLOBAL TECTONICS

Elie de Beaumont (1798–1874), a contemporary of Charles Lyell, was an influential French geologist who took a global view of mountain belts and their tectonics. In his *Coup d'oeil les mines* (a survey of mineral resources) (1824), he gained a knowledge of mountain belts and their mineral resources around the world, including in distant places, such as Africa and Asia.[16] He was aware of Werner's idea that veins of different orientation probably had different ages, and de Beaumont applied this idea to the orientation of mountain belts, which was a fruitful approach to global tectonics and a significant step forward. He was also aware, as Lyell was, that mountain belts could be dated by the unconformity in the strata at their base and the overlying horizontal strata.

Another big picture idea presented by the French geologist was that tectonic revolutions, or mountain-building events, coincided with those extinction events proposed by the anatomist Cuvier, although the number of recognized extinction events and tectonic revolutions were different.[17] This idea today is difficult to dismiss entirely; for example, the assembly of the supercontinent Pangaea destroyed many continental margin ecosystems and may have caused the major Permo-Triassic extinctions, but the idea does not have general application otherwise.[18]

As argued by the historian Mott Greene, de Beaumont's mathematical abilities may have also attracted him to the success of crystallography at that time, whereby crystal morphology could be described quantitatively in terms of integer numbers due the work of the mineralogist René Haüy (who also taught de Beaumont at the Paris School of Mines).[17] De Beaumont unwisely spent much of his subsequent efforts on tectonics in forcing an artificial geometry on mountain belts by fitting great circles on the Earth through mountain belts of different orientation and claiming that a pentagonal geometry resulted. The geometry of these networks became progressively more complex as time went on. An 1863 publication of his was entitled, "Tables of numerical data that secure 159 circles of the pentagonal network," and is more like a treatise on crystallography than tectonics.[19] According to his successor Eduard Suess, Elie de Beaumont's geometrical theories found little appreciation outside France.[20]

One of de Beaumont's major contributions was to take a global view of tectonics, and this tradition was continued in the most important work on tectonics of the late nineteenth century: Eduard Suess's *Face of the Earth* (*Das Antlitz der Erde*), published in four volumes between 1883 and 1888 and translated into English between 1904 and 1909.[20] The book is a synthesis of the existing literature at that time and is dominated mainly by descriptions of the geography, tectonics, stratigraphy, and paleontology of rocks on a global scale. Suess, however, left interpretation to "future generations."[20] Suess follows de Beaumont's lead by focusing on the strike or "trend lines" of mountain belts as being important in correlating mountain belts globally.

In an introductory chapter on dislocations, Suess recognizes that tangential movements cause thrusting and folding in mountain belts, and pays particular attention to the direction of transport (or *vergence* in today's terminology) of fold and thrust belts. The second type of movement he recognized is radial, causing vertical movements. He recognized that the Alps and the Carpathians were thrust northward over the European foreland, and that in the Himalayas the thrusting was in the opposite direction to the south. He attributed this to the different preexisting conditions with regard to the basement rocks. Along with American geologists, he saw the similarities of the Appalachian and Jura fold and thrust belts, and he made a distinction between Atlantic- and Pacific-type coastlines, the latter characterized by coastal mountain belts (what today we would call passive and active margins, respectively). In his second volume, he discusses regressions and transgressions onto the continents, and attributes them to changing sea level, not to continental uplift or sinking. By rejecting regional continental uplift, he implicitly rejected isostasy. The term isostasy was current in the literature by the time Suess published his fourth volume. In volume four, he summarizes the evidence for isostatic compensation beneath mountains and concludes that if a gravity correction is omitted from the calculations, the mass deficiency beneath continents disappears – which, he maintained, was consistent with the geological evidence. Suess can be credited with the name "Gondwána Land," which he applied to the similar fossil remains and stratigraphy of South America, Africa, Madagascar, and the Indian peninsula, providing Alfred Wegner with evidence for his continental drift hypothesis (Chapter 7). He also recognized a larger precursor to the modern Mediterranean, which he named the Tethys Ocean.

ALPINE NAPPE TECTONICS

Suess recalls, in the preface to *Face of the Earth*, meeting Arnold Escher von der Linth (1807–1872), professor at the Polytechnic Institute Zurich in the Swiss Alps: "Escher was a remarkable man. He was one of those of the penetrating eye, which is able to distinguish with precision, amidst all the variety of a mountain landscape, the main lines of its structure. He had just come forward with the magnificent conception, unheard of in the views of that time, of a double folding of certain parts of the Alps, which has since received the name of the double fold of Glarus".[20] When asked why he never published his double fold theory Escher replied "No one would believe me; they would put me into an asylum."[1] Suess probably had the correct answer to his question when he added later: "[B]ut to whom every line he had to publish was a torment." Escher invited Murchinson to see his Glarus structure, and Murchinson confirmed the reality of the structure in his 1848 paper.[7] In this double nappe structure, near horizontal Mesozoic strata overlie folded *younger* Tertiary strata (sandstones and limestones). Moreover, one of the recumbent folds verges to the north and the other verges to the south.

Escher's successor at Zurich, Albert Heim, published the nappe structure in 1878 based on field notes he inherited from Escher (Figure 5.3).[21] The structure is problematic in that the transport directions are in opposite directions. The French Alpine geologist Marcel Bertrand (1847–1907), a student of Elie de Beaumont, noted that the folds in the Tertiary sediments all verge to the north, and reinterpreted the double fold as a single large recumbent fold verging to the north in his 1884 paper (Figure 5.3).[22] This removed the vergence problem, but increased the amount of displacement from 15 kilometers to about 35 kilometers. Such a transport distance was not thought possible as the rocks were likely to crumble due to friction rather than slide. Such large translations were also too much for the contraction theory to account for.[14]

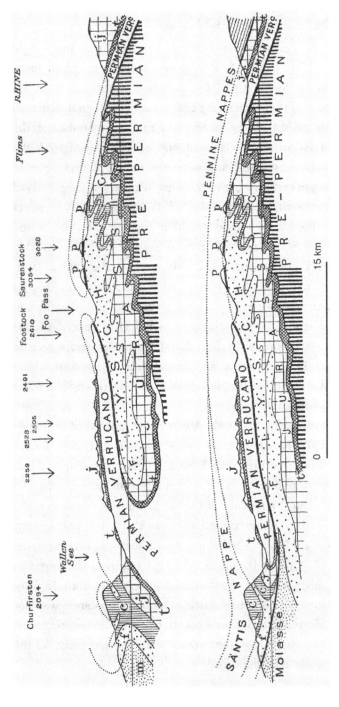

FIGURE 5.3 **Top:** The double fold of Glarus attributed to Arnold Escher, but published by his successor Albert Heim. The two nappes verge in opposite directions and involve Mesozoic strata (Permian, Triassic, and Jurassic) strata emplaced over younger Tertiary folded strata. **Bottom:** A reinterpretation of the double fold as a single north verging nappe by Marcel Bertrand. Note all the folds in the underlying Tertiary (flysch) strata verge north (to the left).

Source: Bailey, E. B., 1935. *Tectonic Essays, Mainly Alpine.* Clarendon Press, London. Courtesy of Clarendon Press.

Bertrand also made a significant contribution to trans-Atlantic geology. He correlated the Caledonian (lower Paleozoic) mountain-building of Scandinavia and Scotland to the northern Appalachians of Newfoundland, and he correlated the Hercynian (middle-to-upper Paleozoic) trend of Europe and the British Isles with the Alleghany mountains of the southern Appalachians.[23] At that time, it was thought that a foundered continent occupied the site of the current Atlantic Ocean and that these mountain structures were present beneath the waves.

Shortly after Bertrand's 1884 paper, Callaway and Lapworth, working independently in the Scottish Highlands, identified thrust faults in which Precambrian gneiss was thrust over lower Paleozoic marine sediments.[24] This thrusting was attributed to the Caledonian mountain-building event and today these structures are referred to as the Arnaboll and Moine thrusts. In the Highlands, nappes were absent, but rather the thrust faults were discrete sliding surfaces. Caledonian belt thrusts were also recognized in Scandinavia where displacements of 100 kilometers were documented. By the end of the century, much of the Prealps (Alpine foothills of France and Switzerland) were shown to be allochthonous (displaced from their source by considerable distance), comprising stacked nappes transported northward.[1] When such nappes were detached from their source (or root zone) as topographically elevated erosional remnants, the term *Klippe* (German for "cliff") was used. The nomenclature in the field of structural geology had become quite complex, and there was no uniformity as to how the terms were used especially in different languages. The work of de Margerie and Heim published in 1888 brought some order to the nomenclature, giving definitions in English, French, and German accompanied by illustrations.[25]

APPALACHIAN FOLD AND THRUST BELT

During the 1830s, fifteen new geologic surveys were established mainly in the eastern United States, all in states close to the Appalachian mountain belt. In North America, the equivalent of William

Smith of England would have been William Maclure (1763–1840),
who single-handedly produced the first geologic map of the United
States east of the Mississippi, in 1809. A Scotsman, he was trained by
Werner in mineralogy and was subsequently successful in business
accumulating wealth as a merchant; he traveled widely throughout
Europe before returning to America.[26] After retiring from business,
he devoted himself to mapping the rocks of the eastern United
States. Following Werner's tradition, he classed his rocks on his
colored map from oldest to youngest as: primitive rocks (in brown),
followed by transition rocks (red), secondary rocks (blue), and alluvial rocks (yellow). His map also shows coal-bearing strata in black,
and his descriptions include locations of many other economic minerals. According to his own account, he crossed the Appalachian
chain fifteen to twenty times at different places along its length,
collecting samples along the way.[27] When his map was published
in 1809, he would have been forty-six years old. His map is certainly
one of the pioneering achievements in American geology by a self-
funded individual at the beginning of the nineteenth century. It was
not until the 1830s that state geologic surveys funded geologic mapping and research.

The father of the Rogers brothers fled Ireland after the rebellion
of 1798. His sons, Henry (1808–1866) and William (1804–1882), studied medicine and chemistry, and became professors at the Universities of Pennsylvania and Virginia, respectively. They were involved
in establishing the Association of American Geologists, which later
became the American Association for the Advancement of Science in
1848, and of which William himself became president in 1876.[28]
Henry and William took positions in the state geologic surveys of
Pennsylvania and Virginia, respectively, and mapped large portions
of those states.

Their 1843 paper on the physical structure of the Appalachian
chain shows they had personal knowledge of Appalachian geology
from New England to Alabama.[29] They described a remarkably coherent mountain chain in excess of 1,000 miles in length and 150 miles

wide in the Valley and Ridge Province (Alleghany Mountains), consisting of a parallel series of anticlines and synclines verging northwest together with thrust faults. Segments of the mountain belt were recognized in map view as alternately convex and concave toward the northwest (promontories and embayments in modern parlance), today interpreted as reflecting the configuration of the original continental margin. In addition they unraveled the stratigraphy of the mountain belt, without error, and relied less on paleontology compared to previous geologists Henry presented his and his brother's results to the Royal Society of Edinburgh in 1856, and his cross-sections (Figure 5.4), which had no vertical exaggeration and were uncommon for the time, came to be widely admired throughout Europe and America as a new standard in geological mapping.[30] In his Edinburgh paper, Henry noted the similarity between the Alleghany Mountains and the Jura of the Alps. The term "Appalachian-style" mountain belt gained currency to reflect the increasing influence of American geological ideas on European thought, which had been dominant up until then.

The Rogers brothers recognized their mountain belt scale observations called for a single cause on a large scale. Although they recognized the role of northwest directed transport because of the asymmetry of their folds, they emphasized vertical motions to explain the anticlines and synclines as due to a wave motion in the underlying liquid substratum (the idea of isostasy would not appear until much later). They say that this idea came from eyewitness accounts of earthquakes where the surface ground undulated in a wave motion. They used the analogy of breaking waves on a beach to explain the northwest vergence of their folds. James Dana later reinterpreted their fold and thrust belt as being due to slow horizontal compression (tangential pressure) produced by long-term cooling of the Earth.

By the early twentieth century, the Valley and Ridge province was interpreted as the result of late Paleozoic horizontal compression along sub-horizontal detachments (mainly along weak shale horizons) with anticline-syncline pairs occurring where the detachment ramps upward to the next younger shale horizon.[31]

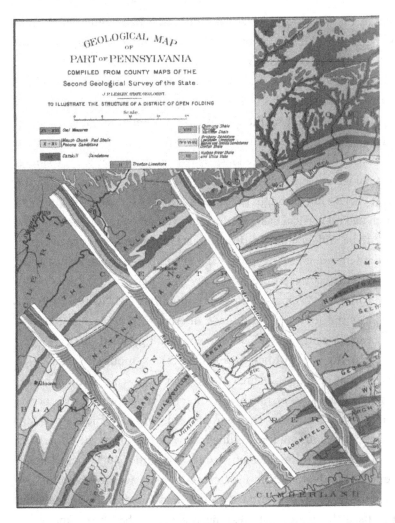

FIGURE 5.4 Geological map of part of Pennsylvania. The mapping and cross-sections are due to H. D. Rogers's work from the 1840s. Many of the cross-sections anticlines show northwest vergence. The Alleghany front is clearly indicated on the map, which today is interpreted to be a major thrust fault.

Source: Willis, B., 1894. The mechanics of Appalachian structure. *United States Geologic Survey, 13th annual report.*

In the late 1950s, the difficulty of pushing brittle thrust sheets from behind due to frictional resistance appeared to be resolved by an important paper by Hubbert and Rubey.[32] They proposed that high fluid pressure along the detachment would substantially reduce frictional resistance and allow thrust belts such as the Jura and Alleghany mountains to be formed. Subsequent experimental work showed that most common rock types had similar frictional properties but that clay-rich rocks such as shale were weaker – this would explain the Alleghany thrust belt where detachments are soled mainly along shale horizons.[33]

THE GEOSYNCLINE CONCEPT

The New York geologic survey was established in 1836 and James Hall (1811–1898) was assigned to investigate the geology and paleontology of the western part of the state where shallowly dipping fossiliferous strata from Ordovician to Mississippian were present. Hall was an extremely productive worker, writing 42 books and 200 articles, mainly on paleontology and stratigraphy.[34] His presidential address to the American Association for the Advancement of Science in 1857 (not published until 1883; possibly suggesting mountain-building was not at the top of his agenda) showed he had a broad perspective on the stratigraphy and paleontology of most of the United States (and parts of Canada), but especially the Appalachian region.[35] He proposed in his presidential address that the Appalachians, and other mountain belts, were formed by thick accumulations of sediments, up to 40,000 feet (about 12 kilometers) in the case of the Appalachians, in an elongated trough or syncline which was subsequently elevated in proportion to the sediment thickness. Because the sediments were deposited in shallow water but were of great thickness, this implied that the trough or syncline deepened as the sediments were deposited and that the weight of the sediments caused the deepening of the trough. As the trough deepened, the strata were compressed and folded and faulted, allowing underlying igneous rocks to infiltrate the strata. Because these sediments thinned rapidly to the west, the idea that the source for the sediments was a landmass to the east of the Appalachians (called Appalachia), now submerged in the Atlantic Ocean, was commonly held at least until 1923.[36]

James Dana, a contemporary of Hall's and a professor at Yale University, (not to be confused with Edward Dana, author of the mineralogy textbook, though both are of Yale and both were editors of the *American Journal of Science*) commented that "in Hall's theory of the origin of mountains, the elevation of the mountains is left out."[37] Nevertheless, Hall's proposal sparked in Dana new ideas on mountain-building, and Dana coined the term geosynclinals, or geosyncline, for Hall's sediment-filled trough.[38] Dana, however, argued instead that the geosyncline was formed not by the weight of the sediments, but by lateral compression as a result of global cooling and contraction (contraction theory). The contraction he argued was concentrated along the continent – ocean boundaries such as the west and east coasts of North America producing the Sierra Nevada and Appalachian mountain chains, respectively. This avoided Dutton's critique, noted already, that global contraction should produce mountain chains of all orientations. Dana further argued that most of the compression came from the oceanic side (because the oceanic crust, he argued, underwent more contraction compared to continental crust), thereby explaining the vergence of folds being toward the interior of the continent on both sides of North America.

The geosynclinal origin of mountains therefore was produced by long continuous subsidence to form a trough that later filled with sediment and finally was compressed laterally to produce a mountain belt by folding and faulting. This hypothesis was closer to a uniformitarian view rather than the catastrophist views of the Rogers brothers. The geosyncline was in effect accreted to the margin of the continent by lateral compression, and Dana strongly hints at continents growing by marginal accretion with time in one of his 1873 papers, particularly with regard to North America, an idea that is still supported today.[39]

John Joly (the Irish geologist we briefly met in Chapter 2 and mentioned earlier in this chapter) considered the effect of radioactivity

on the heating and metamorphism of Dana's Appalachian geosyncline (with 12 kilometers of sediment) in his 1911 book, and concluded that the additional heating would weaken the sediments and
help to mobilize them.[13] The geosyncline concept (which became
increasingly complex in geometry over time) played a major role in
tectonic thought throughout the twentieth century, both in North
America and in Europe, until the development of plate tectonics in
the early 1970s (Figure 5.5).

MYLONITES AND THE ELEVATION OF MOUNTAIN BELTS

The American geologist James Hall can hardly be blamed for leaving
out the elevation of mountains in his theory for mountain belts;
after all, he was primarily a paleontologist and a stratigrapher. The
stacking of nappes and thrust sheets observed in several mountain
belts clearly had a crustal thickening effect, but the mechanics of the
processes were obscure. We saw in Chapter 2 that Charles Lapworth
named the Ordovician period for rocks between the Silurian and
Cambrian based on his work in Wales. In his later work in the
Highlands of Scotland, he coined a second new term for the rocks
found along thrust faults: mylonite (from the Greek *mylon*, a mill).
He stated: "The old planes of schistosity become obliterated, and
new ones are developed; the original crystals are crushed and spread
out, and new secondary minerals, mica and quartz, are developed.
The most intense mechanical metamorphism occurs along the grand
dislocation (thrust) planes."[24] Mylonites are now recognized as an
important rock type that form along major dislocation surfaces in all
mountain belts, whether due to compression or extension; they are
characterized by a fine-grain size and a strong foliation. Understanding the processes involved in their formation is important to understanding how mountain belts are assembled, and accordingly the
modern literature on mylonites is vast.

Although Lapworth noted new minerals of mica and quartz,
implying recrystallization, the subsequent literature on mylonites,

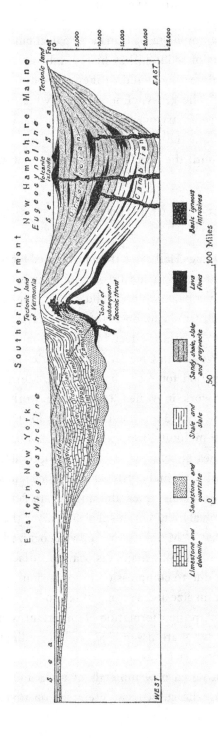

FIGURE 5.5 The northern Appalachian geosyncline concept introduced by Hall and Dana became more complex with time involving a shallow water facies basin (miogeosyncline) and a deeper water facies basin (eugeosyncline) separated by a geoanticline. Eventually, these were interpreted in the late 1950s as corresponding to a back arc basin (miogeosyncline), a volcanic island arc (geoanticline), and a forearc or trench (eugeosyncline).

Source: Kay, M., 1951. North American geosynclines. *Geol. Soc. Am. Mem.*, 48, 48–60.

until relatively recently, has emphasized the cataclastic or brittle nature of the deformation.[40] If mylonites were formed by brittle processes involving frictional sliding, thrust sheets would crumble due to frictional resistance rather than travel long distances intact; this has been the long-recognized paradox of nappes and thrust sheets.

After the discovery that metallurgical principles could explain fine-grained textures in quartz-rich mylonites, it was realized that many mylonites can be explained by plastic rather than brittle deformation.[41] Such plastic or ductile behavior allows a weak detachment surface to develop, allowing recrystallization of quartz and new micas to develop along the foliation at depths of about 10–15 kilometers in the crust. Lapworth was largely correct in his original interpretation, just unfortunate that he chose the mill analogy and implied grinding rather than recrystallization. In the case of the Glarus thrust (Figure 5.3), the mylonites at the contact are derived solely from limestone, which can behave plastically at shallower depths in the crust. Joly, in his 1911 book *Radioactivity and Geology*, considered that the recent discovery of radioactivity as a heat source enhanced the plasticity of nappes at deeper levels and aided in their emplacement.[13]

EXPERIMENTAL STUDIES

In Chapter 4 we noted that a friend of Hutton, Sir James Hall of Scotland, conducted important high-temperature experiments on lavas after Hutton died. Hall's contribution to metamorphism was his conversion of limestone to marble by sealing the sample in a closed steel chamber under high temperature so that volatiles could not escape, mimicking the deeper earth.[42]

In his 1815 paper read to the Society in Edinburgh, Hall described folded outcrops at several locations in coastal Scotland of interbedded slate and limestone.[43] He reproduced similar folds in preliminary experiments using cloth fabrics weighted down by a door ("unhinged at the time") by horizontal blows from a hammer on adjacent vertical wood panels.[43] More exacting experiments were performed using clay as the modeling medium and horizontal screws as the horizontal compressive force. He assumed the inclined strata were originally deposited horizontally, which was a substantial improvement over the Neptunist view that tilted strata were originally deposited in their current position. The Rogers brothers were apparently unaware of Hall's work that folded strata could be produced by horizontal compression; they relied on an underlying fluid-wave action to produce their Appalachian anticlines and synclines as late as the 1840s. By 1888, simple mechanical experiments using plaster of Paris, clay, and sand were producing realistic fold and thrust models.[44] Important modern experiments on ductile[41] and brittle behavior[33] have been mentioned already.[4]

ISOSTASY

During a geodetic survey of India led by Colonel Everest (later Sir Everest), for whom the mountain is named, a discrepancy between trigonometric results and astronomical measurements between two stations 370 miles apart was noticed. The error amounted to 5 seconds of arc (about 0.1 degrees). Everest thought the problem might be the gravitational attraction of the Himalayas on the plume bob used for astronomical measurements. The mountain attraction would pull the plume bob toward the north and lower the latitude estimate, which was what was observed. Everest asked John Pratt, a Cambridge-trained mathematician, then archdeacon of Calcutta, to look more closely at the problem. After exhaustive calculations, Pratt concluded the attraction should be 15 seconds, not 5, so that the observed discrepancy was *less* than expected. Pratt published his results in 1855, but had no explanation for the discrepancy and concluded: "The whole subject, however, deserves careful examination; as no anomaly should remain unexplained in a work conducted with such care, labour and ability."[45]

Pratt did not have long to wait because in the same issue of the journal, the following paper was authored by G. B. Airy, the Royal Astronomer in Greenwich (who had reviewed Pratt's paper). In a short

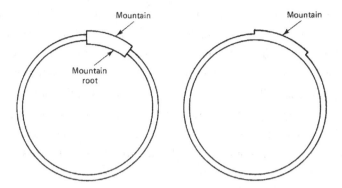

FIGURE 5.6 The left-hand side illustrates a mountain belt as envisioned by Airy in which the mountain is in hydrostatic equilibrium with the underlying mantle – which is considered to be a viscous fluid. The gravitational attraction of this mountain would be less than that shown on the right-hand side, which was the situation that Pratt considered. In this case, the crust is strong enough to support the mountain belt. The Airy interpretation is the more correct one.

Source: Airy, G. B., 1855. On the computation of the effect of the attraction of mountain-masses, as disturbing the apparent astronomical latitude of stations in geodetic surveys. *Phil. Trans. Roy. Soc. London*, 145, 101–104.

non-mathematical paper, he suggested a resolution to the problem that provided important new insight into how the continental crust behaves.[46] He suggested that mountains were floating on a liquid substratum analogous to a floating log. Mountains had "roots" which displaced the denser substratum (today's mantle) with less dense continental material. This is the same as Archimedes' principle whereby the mountains are buoyed up by a force equal to the weight of displaced mantle. The result is that there is less gravitational attraction than expected, especially when compared to a mountain that has no roots (Figure 5.6). It was not until 1889 (thirty-four years after Airy's paper) that the American geologist Clarence Dutton named the process of continental hydrostatic equilibrium isostasy, and he outlined the geological implications of the process: namely, denudation would cause mountains to be uplifted vertically until they reached equilibrium again, and conversely deposited sediments would download the adjacent crust.[47]

In Chapter 2 we saw the success of stratigraphy and development of a geologic timescale, which was a major achievement for geology as a fledgling science. Toward the end of the nineteenth century, however, tectonic problems were beginning to accumulate to the point where disarray was apparent. Dutton's outburst on contraction theory earlier in the chapter (that the theory "should be thrown overboard") is one example, and Dana's comment on Hall's ideas (that it was a theory of mountains with the mountains left out) suggested that things were not quite right in the tectonic world.

An important implication of isostasy was that continents could not founder and become the sites of ocean basins as was commonly thought at the time. This raised serious problems for trans-Atlantic geologic correlations, such as those proposed by Bertrand and Suess involving correlations of fossils and stratigraphy across the North and South Atlantic and between Africa and India. The discovery of radioactivity (Chapter 2) essentially invalidated the contraction theory as due to long-term cooling, although Jeffreys did revive the theory, by including heating due to radioactivity, so that it survived into the twentieth century.[15] Large horizontal displacements demonstrated for Alpine nappes and thrust sheets were mechanically highly problematic. The Rogers brothers explained the Alleghany Mountains as catastrophic in nature (i.e., earthquake-like) at a time when Lyellian uniformity was the more widespread doctrine throughout geology. There was clearly a crisis in tectonics, particularly with regard to the origin of mountain belts.

World War I and World War II interrupted scientific progress in the first half of the twentieth century, prolonging the crisis. Continental Drift, proposed in 1912, was largely rejected by geologists until the late 1950s, and the geologic crisis which had its origin in the late nineteenth century would not be resolved until almost a century later, in the 1970s. In the next chapter we look at Alfred Wegener's hypothesis and why it was largely rejected, with few exceptions, in both Europe and North America.

REFERENCES

1. Bailey, E. B., 1935. *Tectonic Essays, Mainly Alpine.* Clarendon Press, London.
2. Descartes, René, 1644. *Principles of Philosophy.* Trans., V. Miller and R. Miller, 1983. Riedel Publishing, Dordrecht.
3. Leibniz, G., 1669. *Protogaea.* Trans., C. Cohen and A. Wakefield, 2008. Chicago University Press, Chicago.
4. Mitchell, J., 1760. Conjectures concerning the cause and observations upon the phenomena of earthquakes. *Philosophical Transactions,* v. 55, 566–634.
5. von Zittel, K. A., 1901. *History of Geology and Paleontology to the End of the Nineteenth Century.* Trans., by Ogilvie-Gordon. Scott, London.
6. Murchinson, R. 1841. First sketch of some of the principal results of a second geologic survey of Russia. *Philosophical Magazine,* v. 19, 417–422.
7. Murchinson, R. 1848. On the geologic structure of the Alps, Apennines and Carpathians. *Proceedings of the Geological Society of London,* v. 5 157–312.
8. Lyell, C. 1830–1833 (1997). *Principles of Geology (3 vols.).* Penguin Books, London.
9. Le Conte, J., 1878. On the structure and origin of mountains with special reference to the recent objections to the contractional theory. *American Journal of Science,* v. 16, 95–111.
10. Fisher, O., 1882. Physics of the Earth's crust. *Nature,* v. 27, 76–77.
11. Fisher, O., 1888. On the mean height of the surface elevations of a solid globe through cooling. *Philosophical Magazine,* v. 25, 7–20.
12. Dutton, C. E. 1874. A criticism upon the contractional hypothesis. *American Journal Science,* 3rd series, v. 8, 113–123.
13. Joly, J., 1911. *Radioactivity and Geology: An Account of the Influence of Radioactive Energy on Terrestrial History.* Constable & Co, London.
14. Holmes, A. 1925. Radioactivity and the Earth's thermal history part IV: A criticism of parts I, II and III. Geological Magazine, v. 62, 504 – 515.
15. Jeffreys, H. 1924. *The Earth: It's Origin, History and Physical Constitution,* 1st ed. Cambridge University Press, Cambridge.
16. Elie de Beaumont, L., 1824. *Coup d'oeil sur les mines.* Levrault, Paris.
17. Greene, M. T., 1982. *Geology in the Nineteenth Century.* Cornell University Press, Ithaca.
18. Gradstein, F., Ogg, J., and Smith, A., 2004. *A Geologic Time Scale.* Cambridge University Press, Cambridge.
19. Elie de Beaumont, L. 1863. Tablieu des données numériques qui fixent 159 cercles du reseau pentagonal. *Comptes Rendus des séances de l'académie des sciences,* v. 57, 1–12.

20. Suess, E., 1904–1909. *The Face of the Earth (4 vols.)* Clarendon Press, Oxford. Trans., by H. Sollas. First published as *Das Antlitz der Erde,* 1883–1904.

21. Renevier, E. 1879. Notice on Professor Heim's work on the mechanisms of mountains. *Geological Magazine,* v. 6, 131–135.

22. Bertrand, M., 1884. Rapport de structure de Glaris et du basin houiller du nord. *Bulletin Société Geologique France,* series 3, v. 12, 318–330.

23. Bertrand, M., 1887. Le chaine des Alpes, et la formation du continent Européen. *Bulletin Société Geologique France,* series. 3, v. 15, 423–447.

24. Lapworth, C., 1885. The Highlands controversy in British geology; its causes, course and consequences. *Nature,* v. 32, 558–559.

25. De Margerie, E. and Heim, A., 1888. *Les dislocations de l'écorce terrestre. Die dislocationen der erdrinde. Essai de définition et de nomenclature.* Verlag von Wurster, Zurich.

26. Merrill, G. P., 1924. *The First Hundred Years of American geology.* Yale University Press, New Haven.

27. Maclure, W., 1809. Observations on the geology of the United States, explanatory of a geological map. *Transactions of American Philosophical Society,* v. 6, 411–428.

28. Rodgers, J., 2008. *Rogers, H. D. Complete Dictionary of Scientific Biography.* Charles Scribner's Sons, Detroit, 504–506.

29. Rogers, W. B. and Rogers, H. D., 1843. On the physical structure of the Appalachian chain, as exemplifying the laws which have regulated the elevation of great mountain chains, generally. *Association of American Geologists and Naturalists.,* 474–503.

30. Rogers, H. D., 1856. On the laws of structure of the more disturbed zones of the Earth's crust. Transactions Royal Society *Edinburgh,* v. 21, 431–470.

31. Rich, J. L. 1934. Mechanics of low-angle overthrust faulting as illustrated by the Cumberland Mountain thrust, Virginia, Kentucky and Tennessee. *American Association Petroleum Geologists Bulletin* v. 18, 1584–1596.

32. Hubbert, M. and Rubey, W., 1959. Role of fluid pressure in the mechanics of overthrust faulting. *Geological Society of America Bulletin,* v. 70, 115–166.

33. Byerlee, J. D., 1978. Friction of rocks. *Pure Applied Geophysics,* v. 116, 615–626.

34. Dott, R. H. Jr., 2005. *James Hall (1811–1898), A Biographical Memoir,* v. 87, 1–19. National Academies Press, Washington DC.

35. Hall, H. D., 1883 (1857). Contributions to the geological history of the American continent. *Proceedings of the American Association for the Advancement of Science* v. 31, 29–69.

36. Schuchert, C., 1923. Sites and nature of American geosynclines. *Bulletin of Geological Society of America*, v. 34, 151–229.

37. Dana, J., 1866. Observations on the origin of some of the Earth's features. *American Journal Science*, 2nd series., v. 42, 205–211.

38. Dana, J. 1873. On some results of the Earth's contraction from cooling, including a discussion of the origin of mountains and the nature of the Earth's interior. Part I. *American Journal of Science*, 3rd series., v. 5, 423–443.

39. Dana, J., 1873. On some results of the Earth's contraction from cooling, including a discussion of the origin of mountains and the nature of the Earth's interior. Part II. The condition of the Earth's interior and the connection of the facts with mountain making. Part III Metamorphism. *American Journal of Science*, v. 6, 31–36.

40. Higgins, M. W., 1971. Cataclastic rocks. *U.S. Geol. Survey*, Professional Paper 687, 1–97.

41. Griggs, D. T. and Blacic, J. D., 1965. Anomalous weakness of synthetic crystals. *Science*, v. 147, 292–295.

42. Hall, J. Sir, 1805. Account of a series of experiments showing the effect of compression in modifying the action of heat. *Transactions of Royal Society of Edinburgh.*, v. 6, 71– 185.

43. Hall, J. Sir, 1815. On the vertical position and convolutions of certain strata and their relation with granite. *Transactions of Royal Society of Edinburgh*, v. 7, 79–108.

44. Cadell, H. M., 1888. Experimental researches in mountain-building. *Transactions of Royal Society of Edinburgh*, v. 35, 337–357.

45. Pratt, J. H., 1855. On the attraction of the Himalaya mountains and of the elevated regions beyond them upon the plumb-line in India. *Philosophical Transactions of Royal Society of London*, v. 145, 53–100.

46. Airy, G. B., 1855. On the computation of the effect of the attraction of mountain-masses, as disturbing the apparent astronomical latitude of stations in geodetic surveys. *Philosophical Transactions of Royal Society of London*, 145, 101–104.

47. Dutton, C. E., 1889. On some of the physical problems in geology. *Bulletin of Philosophical Society of Washington*, v. 11, 51–64.

6   Continental Drift

Thus the theory of continental drift is a fairytale.

– Bailey Willis, 1944[23]

As early as the 1870s, geologists working in Africa[1] and India[2] had concluded there was once continuity between Africa and India, and also Australia, based on the similarity of plant and reptile fossils on these three continents in Permo-Carboniferous times. At the time it was commonly assumed that the ocean basins were simply foundered continents, so it seemed plausible that continental connections once existed between now separate continents. Recall, for instance, from Chapter 5 that the French geologist Bertrand correlated mountain belts across the north Atlantic Ocean from Europe to North America believing that these same mountain belts underlay the present day Atlantic; the fact that continental crust and oceanic crust were of entirely different chemical compositions was not yet fully appreciated. As an alternative to foundering continents, Darwin's coral reef map of 1842 suggested to others that there were shallow underwater mountains throughout the Indian Ocean (Maldives, Seychelles, etc.), so that a land bridge may have existed between India and southern Africa.[2]

Edward Suess, in his *Face of the Earth*, coined the term *Gondwanaland* to describe this southern continent, named after Permian to Jurassic formations from India (his Gondwanaland also included South America east of the Andes and Madagascar). He stated that because *Glossopteris* (a Permian-aged fern) was common to all the Gondwana continents and that all its Permian sediments were non-marine, that it was a contiguous continent at that time and only began to break up with the appearance of Jurassic marine sediments. Suess rarely ventured into tectonic mechanisms and, apart from subsidence of ocean basins, he left them vague.

In Chapter 5 we saw that horizontal thrust and nappe displacements on the order of tens of kilometers (up to 100 kilometers in the case of the Caledonian of Scandinavia) were by now widely accepted, but large-scale horizontal displacements of the continents themselves had not yet been seriously proposed. This changed at the beginning of the twentieth century with the publication of a paper by the American geologist Frank B. Taylor in 1910,[3] followed by Alfred Wegener's book on Continental Drift in 1915.[4] Taylor's paper might be termed Continental Drift light: he allowed for continental displacements of hundreds of kilometers, whereas Wegner's hypothesis called for horizontal displacements of thousands of kilometers.[5]

In the *Face of the Earth*, Suess had used trend lines, or map-view trends of mountain belts and island arcs, to analyze tectonic patterns of Asian mountain belts. In Asia he recognized a series of arcs that are convex toward the Pacific Ocean, and he interpreted them as indicating tectonic transport toward the south. In the case of the Himalayas, he saw that the Indian continent acted as a rigid indenter causing a reentrant into Asia, a geometry that is visible on any physical geography map of the region. Taylor applied this approach of using tectonic trends to Tertiary mountains belts in Europe and also North America. He concluded that all of the northern hemisphere continents were transported southward to lower latitudes (despite the fact that many of the mountain belts showed northward tectonic transport, such as the Alps and the Carpathians) or eastward transport (the Rocky Mountains). He then applied the same method to the southern hemisphere, where he concluded that South America and Australia indicated tectonic transport to the north to lower latitudes (despite the fact that the Andes indicate tectonic transport to the west). He interpreted the tectonic displacements in the two hemispheres as due to flattening of the Earth at the poles, which gave the Earth its oblate shape, thereby causing the continental lithosphere to move toward the equator. Although his tectonic analysis does not stand up to even cursory scrutiny, he did make some original suggestions regarding rifting along the mid-Atlantic ridge and suggested that it was the site along which Africa and South America fragmented.

THE ORIGIN OF CONTINENTS AND OCEANS

Alfred Wegener was born in Berlin in 1880 and died in 1930 on his third expedition to Greenland.[6] He received a Ph.D. in astronomy in Berlin in 1905, but spent most of his career studying meteorology and the physics of the atmosphere. On his first expedition to Greenland (1906–1908), he undertook atmospheric investigations using balloons and kites. On return from Greenland he took a teaching job in meteorology at the Physical Institute at Marburg, Germany, and from 1909 to 1912 he published more than forty papers on atmospheric physics that he later collected into a book.

While examining a new atlas of the Atlantic in 1910 based on data collected on the Challenger voyage (1873–1876), Wegener noticed the similar bathymetry on the opposite coastlines of Africa and South America. As he explains in the introduction to his book, the following year he came across geological and paleontological data showing that the two continents were probably connected at one time as Suess and others before him had suggested. In 1912 he wrote a paper on his theory of horizontal displacement of the continents (later called Continental Drift). During a prolonged sick leave from action in World War I, he wrote the book for which he is now most famous: *The Origin of Continents and Oceans* in 1915. The war meant it did not receive the attention it deserved, and it was not translated into English and other languages until 1924 (from the third German edition), after which it aroused great interest worldwide.[5] A fourth edition was published in 1929 in which Wegener placed greater emphasis on geodetic measurements of present day continental movements. What follows is a brief outline of Wegener's book (first English edition).

The first chapter of Wegener's book is a brief synopsis of the hypothesis of horizontal displacement of the continents, what Wegener called his "displacement theory" (Figure 6.1). Chapter 5 in this book summarized why the contraction theory of mountain-building is

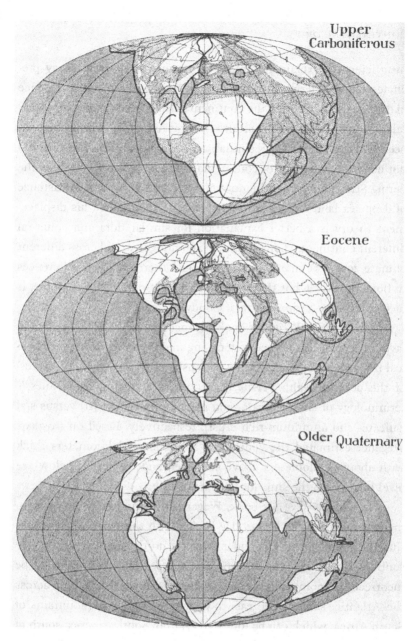

FIGURE 6.1 Reconstruction of the world geography according to Wegener for three periods: upper Carboniferous (~290 million years ago), Eocene (~45), and lower Quaternary (~2). Stippled areas are shallow continental seas. Continental outlines include the submarine continental shelf. Wegener called the Carboniferous supercontinent Pangaea. The Atlantic Ocean is still quite narrow by Eocene time. The north Atlantic is still nearly closed by the lower Quaternary, making the north Atlantic basin much too young.

Source: Wegener, A. 1924. *The Origin of the Continents and Oceans.* Methuen, London. Trans., J. G. A Skerl. First published in German 1915.

insufficient to explain mountain shortening and why isostasy precludes the foundering of continents into ocean basins. The absence of deep sea sediments on the continents also argued against continental foundering and the only alternative was narrow land-bridges across oceans to explain the fossil similarities on opposite sides. Wegener did not deny the existence of continental shelf land bridges (such as the Bering Straits) along continental margins, but did deny the existence of deep sea land bridges across oceans and maintained his displacement theory is a better explanation for similar flora and fauna on different continents. Long distance land bridges would cross different climate zones, he maintained, and would therefore show differences in flora and fauna; but the Permian fern *Glossopteris*, for example, is identical on the various parts of Gondwanaland.

Chapter 3 of Wegner's book advances geophysical arguments in favor of displacement theory, focusing mainly on the different physical properties (seismic, magnetic, density, and topographic features) of the oceanic realm versus the continental realm, using Suess's terminology of sima (silicate- and magnesium-rich crust) versus sial (silicate- and aluminum-rich crust), respectively. Based on isostasy, Wegener estimated the continents are about 100 kilometers thick with about 5 kilometers above sea level, and the rest is below sea level floating on the sima (oceanic) substrate.

In Chapter 4 of his book, Wegener presents the geologic arguments in favor of his theory. He begins with the Atlantic and notes that the south Atlantic is wider than the north Atlantic, so that the south Atlantic was the first to open up (we now know this to be incorrect). He points out various geologic features that join up across the Atlantic, such as the east-to-west trending Cape Mountains of South Africa, which can be followed again in South America south of Buenos Aires, and that the Precambrian gneisses of the Brazilian shield match gneisses on the west coast of Africa. Moving northward, he summarizes Bertrand's correlation of older mountain belts across the North Atlantic with Europe (see Chapter 5). Wegner uses the analogy of torn newspaper fragments matching up with each other for the opposite sides of the Atlantic and the various geologic features, such as older mountain belts also lining up, so that not only did the shape of the torn fragments align, but the lines of print could be read across the torn newspaper.

In the north Atlantic, Wegner follows Taylor's ideas on the rifting of Greenland from Labrador in the west and from Norway to the east. Turning to connections between India, Madagascar, and Africa, Wegener summarizes the evidence presented earlier by Suess for the existence of Gondwanaland. In Chapter 5 he summarizes a previously published survey of twenty paleontologists regarding the timing of the disappearance of land bridges based on the paleontological evidence. The paleontological data indicate a breakup of Gondwanaland between the Jurassic and Cretaceous. However, Wegener apparently over-interprets the data in the case of Europe and North America where he concludes a Quaternary breakup is indicated, when in fact most of the data indicated a Jurassic breakup. This mistake explains why Wegener includes, in his first chapter, a correlation of terminal glacial moraines in Europe and North America as geological evidence of a connection as late as the recent Pleistocene glaciations (this was a major mistake). It also explains why in his reconstruction the geography of today does not occur until the Quaternary period only a few million years ago (see Figure 6.1), when in fact our present-day geography becomes quite recognizable by the late Cretaceous, 65 million years ago. These mistakes played a large role in the criticisms of his theory as it implied the Atlantic Ocean was much younger than the Pacific Ocean, which many geologists were not ready to accept.

In Chapter 6, Wegener summarizes paleoclimatic arguments. The glacial deposits (tillites) in the various parts of Gondwana are strong evidence that Gondwanaland was centered over the South Pole in Carboniferous time. The Permian fern *Glossopteris* generally overlies these glacial formations. At the same time in equatorial regions, tropical coals and evaporate deposits are present and these observations do indeed support the existence of Gondwanaland. In Chapter 7 of his book, Wegener attempts to present geodetic measurements based on astronomical measurement of longitude to indicate present day rates of drift, but the method is not sufficiently accurate to provide an answer. This chapter is considered to be Wegener's weakest. In the fourth edition of his book, he provides additional geodetic data, but most commentators gave them little credence.

The final chapter discusses the driving forces for drift. Wegener saw a westward drift of the continents (e.g., South America) and, like Taylor before him, he also saw a drift toward the equator. He attributed the westward drift to tidal forces of the sun and moon, and the equatorial drift as due to the Earth's equatorial bulge. The Cambridge geophysicist Harold Jeffreys (1891–1989) showed these forces to be hopelessly inadequate to move the continents (see Box 6.1).[7]

The possible reasons for the rejection of Continental Drift are many and may have included Wegener's nationality at the time of World War I, as well as his outsider status as a nongeologist. Continental Drift was more strongly rejected in America compared to Europe. The exhaustively researched book *The Rejection of Continental Drift* by Naomi Oreskes specifically addresses this differential response.[8] She concludes that American geologists compared to Europeans had a different view of how science should be practiced. A simpler explanation, however, is that most of the evidence in favor of drift was on the continents in the southern hemisphere, with which most American geologists (excepting Reginald A. Daly) were unfamiliar, but which European geologists were more familiar.

The structure of Wegener's book itself may have been also part of the problem; the first chapter presents the hypothesis and the subsequent chapters provide the supporting evidence, which may have raised the ire of some observers. The influential American geologist Thomas Chamberlin (1843–1928), who is best known for his studies of Pleistocene glaciations, published an influential paper entitled "The Method of Multiple Working Hypotheses"; the paper has no references, a testament to its originality.[9] This paper is still influential in the United States, and it was reprinted in 1966 by

BOX 6.1 **Geophysicist Harold Jeffreys (1891–1989)**

Because of his fame as an authoritative geophysicist, Harold Jeffreys's opinions against drift were given great weight by many. Jeffreys was born in Durham, England, and died in Cambridge, aged ninety-eight. He attended Durham University (now University of Newcastle) where he graduated in 1910 with a first-class degree with distinction in mathematics; he also studied chemistry and physics, plus a year of geology.[31] He was awarded a scholarship to read mathematics at Cambridge, where he received first-class marks, and his scholarship was extended to four years while he developed research focused on astrophysics. In 1922 he returned to Cambridge as a lecturer in mathematics. While in London he met Arthur Holmes when Holmes was working on radiometric dating, and Jeffreys then became interested in the age of the Earth and its thermal history and other geophysical problems. The results of these studies were brought together in his great treatise *The Earth: Its Origin, History and Physical Constitution* (1924), which ran six editions The sixth edition lists 170 publications of his and is a small fraction of his total output in other fields.[32] His main achievements lie in the fields of seismology, planetary geodynamics, meteorology, applied statistics, and Earth's gravity anomalies. He showed, for example, from seismology that the Earth's metallic outer core is liquid, leading to new ideas concerning the origin of the Earth's magnetic field. However, in geology, he was on the wrong side of several major debates.

As outlined in Chapter 5, the idea that long-term cooling of the Earth leads to contraction of the crust and formation of mountains chains (contraction theory) was dismissed as inadequate on quantitative grounds. Jeffreys, however, revived (resurrected might be a more appropriate word) the theory in his book *The Earth* (second edition) by taking into account radioactivity and pronounced the shortening as being adequate (100–400 kilometers) to explain Tertiary mountain belts. In answering objections to the theory, he said, with detectable ire: "Some objections have been answered several times already, but appear to be capable of indefinite

BOX 6.1 **(cont.)**

repetition however often they are answered."[7] He did not answer a major objection to contraction theory, however, namely that it could not explain the asymmetry of mountain belts. Although in the sixth edition, where he still supported contraction theory, he invoked James Dana's old theory of mountain-building, which attempted to explain the asymmetry.

Jeffreys was not a fan of continental land bridges because isostasy precluded their foundering (for once, here he agreed with Wegener). He noted that many were opting for Continental Drift as an alternative, and commented: "If ever there was a migration from the frying pan into the fire it is this."[7] His two main geophysical arguments against drift in the second and sixth editions of his book are the same: lack of a driving force, and the strength of the sima (oceanic crust) would not allow the sial (continental crust) to drift through it. These ideas are summarized throughout this chapter, and are not repeated here. Jeffreys also rejected that the sima was a viscous fluid as Wegener proposed – the topography of the ocean basins would have decayed to become a flat surface, which in 1976 was known to have highly variable topography.[32]

Regarding convection models, he complains that the rheology is not specified in the various models and therefore they cannot be evaluated, and this presumably applies to Holmes's 1929 paper. He concluded convection would lead to a steady state rather than episodic mountain-building events. Regarding the paleomagnetic results that had been accumulating since the 1950s that showed large continental displacements, he was suspicious of the stability of the magnetic field in minerals, saying: "I have been told that the magnetic minerals, magnetite and hematite, stand ill-treatment better than steel. But I remain doubtful."[32] By 1976 (sixth edition of *The Earth*), the basic rudiments of plate tectonics were in place, but Jeffreys rejected both sea-floor spreading and the concept of subduction, saying that the mantle was too strong to digest the slab. At this time he was eighty-five years old. He published his last

BOX 6.1 **(cont.)**

technical publication at age ninety-six. It might be said that Harold Jeffreys was to geologists of the twentieth century what Kelvin was to nineteenth century geologists: a thorn in their sides. Recall that Kelvin was also on the wrong side of several major geological debates, including Darwin's evolution and the age of the Earth.

Science Magazine as a tribute to its importance. After its original publication, it may have influenced the reception to scientific hypotheses in general, and more specifically, Wegener's drift hypothesis. Chamberlin recognized three intellectual states: the ruling theory, the working hypothesis, and the method of multiple working hypotheses. Wegener, in his book, by stating his hypothesis upfront in the first chapter followed by supporting evidence, may have alienated the American scientific community at that time who favored Chamberlin's multiple hypothesis approach. Several critics noted that Wegener was an advocate for his own hypothesis rather than an impartial investigator, and that his was an unscientific approach; this argument is consistent with Oreskes's conclusion that American and European scientists had different approaches to science.[8] Wegener did, however, have a few influential supporters.

WEGENER'S SUPPORTERS

Three early and strong supporters of Wegener were the English geologist Arthur Holmes, the South African geologist Alexander du Toit, and Émille Argand, a Swiss Alpine geologist. The American igneous petrologist Reginald A. Daly and John Joly of Ireland also supported their own versions of Drift.

Arthur Holmes, who we met in Chapter 2 for his work on the geologic timescale and in Chapter 3 for his work on the age of the Earth, was a strong advocate of convection in the mantle as the driving force for Continental Drift as early as 1929. Holmes was born in Tyne,

England, in 1890, and died in London in 1965.[10] He gained an interest in geology in high school and entered Imperial College London and studied physics for his first degree under Robert Strutt (later Lord Rayleigh). He later studied geology under W. Watts at Imperial College. He undertook graduate studies under Strutt based on the recently developed uranium-lead method of dating of minerals, and produced the first geologic timescale based on this work at the age of twenty-one (Chapter 2). Two years later, his book *The Age of the Earth* appeared.[11] He later went on expedition to Mozambique, where he gained field experience and studied mainly igneous rocks. He took a teaching job at Imperial College during the period 1912–1920, then became chief geologist for an oil company in Burma (now Myanmar) from 1920 to 1924, after which he became professor of geology at the University of Durham in 1925. He married for the second time in 1939 to a well-known geologist at that time, Doris Reynolds. He became chair of geology at the University of Edinburgh in 1943. He started writing his textbook *Principles of Physical Geology* while at Durham during World War II, and it was published in 1945.[12] The final chapter is a review of Wegner's Continental Drift theory, where he points out some of its errors but nevertheless acts as a strong supporter of the hypothesis; he also restates convection as the driving mechanism for Continental Drift. The second edition of his *Principles* was published in 1964, a year before his death, which ran to nearly 1,300 pages and covered a wide range of disciplines from geophysics to geomorphology. It became very popular, and was also my first undergraduate geology textbook at college and certainly an intimidating one. His writing was clear and lucid, and he had skill at illustrating his ideas. Holmes received many awards, including the Penrose Medal from the Geological Society of America, its most prestigious award, on his retirement in 1956. His main contributions to the Earth sciences were in geochronology, the geologic timescale, the origin of igneous rocks, and convection in the mantle.

Holmes's 1929 paper entitled "Radioactivity and Earth Movements" is remarkable for how close it comes to modern plate tectonics

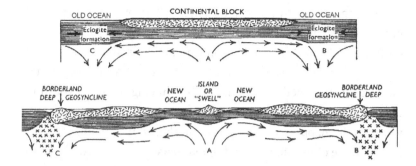

FIGURE 6.2 **Upper panel:** Convection currents in the mantle rise beneath continental crust (A) and down-going currents convert basalt (horizontal lines) to heavier eclogite (B and C). **Lower panel:** Continent is thinned and extended, creating new ocean between continental fragments (A). The eclogite joins down-going convection currents at the edge of the continents, forming borderland geosyncline deeps.
Source: Holmes, A., 1929. Radioactivity and Earth movements. Transactions Geological Society Glasgow, v. 18, 559–606. Courtesy Geological Society of London.

(Figure 6.2).[13] In it he shows how radioactivity is a sufficient energy source for convection to occur in the mafic substratum (the mantle), upon which the continents float isostatically. He assumed the mafic substratum to be a viscous fluid with no yield strength. Upwelling convection currents caused continents to drift apart, forming new basaltic ocean crust, and down-going limbs of convection currents caused continent collisions, formation of geosynclines, and compressional mountain belts (Figure 6.2). He estimated the continents moved at about 5 cm/year, which is the correct order of magnitude. Convection in the mantle provided a driving mechanism for Wegener's Continental Drift, and the lack of strength of the mantle allowed the continents to plow through the mafic substratum, thereby removing two of the chief objections to Wegner's ideas. Collisional mountain belts were produced by converging convection currents. Perhaps because Holmes's arguments were only semi-quantitative, they did not gain substantial support.

Émile Argand, another supporter of Wegner, was born in Geneva in 1879 and died in Neuchâtel in 1940.[14] Because of his artistic skills, his father apprenticed him to an architect, but his mother wanted him to study medicine. While at the University of Lausanne, he met the Alpine geologist Maurice Lugeon, and under his direction Argand decided to devote himself to geology, specifically Alpine structural geology. He was a good mountain climber, having spent his youth in Geneva in the foothills of the Alps. He unraveled the structure of several Alpine nappes, and his excellent ability at drawing complex shapes in three dimensions from different perspectives allowed him to present accurate cross-sections of these structures.[15] The latter publication contains thirteen cross-sections on the same page, showing the development of the nappes of the western Alps from the original paleo-geography through the various tectonic stages, and is a work of art in itself apart from its scientific significance. He was appointed professor of geology at Neuchâtel in 1911. Argand read Wegener's book in the original German in 1915 (during World War I it was illegal to read German in private or in public in Switzerland),[16] and was convinced of the "mobilist" view of the Earth, a term he coined himself. Figure 6.3 shows the complex nappe structure of the Alps due to collision of Africa with Europe, in keeping with Wegner's drift hypothesis. Argand's most important work is his *La Tectonique de l'Asie*, (The Tectonics of Asia), published in 1922.[16] The title is somewhat misleading as the book covers not just the tectonics of Asia, but also of Europe; it was translated into English in 1977.[17] Argand knew at least six languages, and whenever his review of the world literature required a new language he learned that language, sometimes in as little as a few days.[17] Argand's work was influential in French-speaking Europe, but less so in the United States.

Alexander du Toit was born in South Africa (1878–1948).[18] The American geologist Reginald Daly called him the "world's greatest field geologist."[18] It is estimated he mapped 256,000 kilometers (100,000 square miles) during his lifetime using a plane table with a bicycle for transport. He graduated from the University of Cape Town and spent two years studying mining engineering in Glasgow, Scotland, where he met his wife, and he also studied geology at the Royal

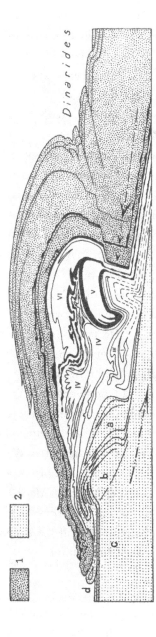

FIGURE 6.3 Diagram showing the development of nappes in the western Alps due to the collision and thrusting of Africa over Europe. 1: Africa; 2: Europe. Mafic rocks are shown in black. Nappes: IV, Great Saint Bernard; V, Monte Rosa; VI, Dent blanche. Scale: 1:1,000,000.

Source: Argand, É. 1922. La tectonique de l'Asie. Conférence faite á Bruxelles, le 10 août. Congres géologique international, Belgique.

College, London. He became lecturer at both the Royal Technical College, Glasgow, and the University of Glasgow. In 1903 he returned to South Africa where he was a field geologist for the Geological Commission of the Cape of Good Hope and spent nearly all of his time in the field over the next seventeen years. In 1923 he received sponsorship from the Carnegie Institution of Washington (USA) to compare the geology of South America with South Africa, with a focus on testing Wegener's hypothesis.

He spent five months studying the geology of Brazil, Paraguay, and Argentina traveling by train and steamer, hosted by the various heads in charge of geological resources of their respective countries which cover a vast area on a continental scale. Not surprisingly, he noted little coordination between countries over this large continent and his synthesis of South American geology is a triumph. The results of his work were published in *A Geological Comparison of South America with South Africa* in 1927, in which he concluded the numerous similarities between the two continents favored the theory of Continental Drift.[19] He published his well known-book *Our Wandering Continents: A Hypothesis of Continental Drifting* in 1937, in which he displays his knowledge of global geology and paleontology, especially that of Gondwanaland.[20]

One of the few well-known American geologists (born in Canada) to support Continental Drift was Reginald Daly. In his 1926 book, *Our Mobile Earth*, he largely accepted Wegner's ideas and proposed the driving mechanism was gravity whereby continents slide off topographic highs on the Earth's surface,[21] which is somewhat similar to Joly's model, which is outlined later in the chapter. Daly also played a role in helping du Toit get support for his South American trip from the Carnegie Institution.

CRITICS OF DRIFT

Petroleum geologists were very interested in the Wegener hypothesis; the breakup of Gondwanaland in Jurassic to Cretaceous times created new passive continental margins where thick, carbon-rich sediments were likely to develop into hydrocarbon reservoirs. Today we know that most of the giant oil fields discovered in the 1980s and 1990s were formed in this tectonic setting. So it is not surprising that the American Association of Petroleum Geologists convened a symposium in New York in February of 1928 to discuss Continental Drift.

The international meeting was convened by van Waterschoot van der Gracht, vice president of Marland Oil Company at that time, and the results of the meeting were published by the University of Chicago Press.[22] The vice president provided an informed introduction to the volume and at the end he addressed the concerns of those who objected to Wegener's Drift hypothesis. Of fourteen participants in the Petroleum Geologist's meeting, four were European and the remainder American. Alfred Wegener's contribution was short and was presented by a surrogate; Wegener himself was preparing for his third and last trip to Greenland. Frank Taylor was a Wegener supporter and largely repeated his 1910 paper described above. Seven Americans were *against* and only one was *for* the Continental Drift hypothesis.

Among the Americans who rejected drift were world-class geologists Bailey Willis of Stanford University; Thomas Chamberlin of the University of Chicago; and Charles Schuchert and C. H. Longwell, both of Yale University. John W. Gregory of the University of Glasgow, who wrote a book on the tectonics of Asia himself,[23] was also against drift. One critic, E. Berry (University of Baltimore) called the hypothesis unscientific because Wegener was selective in his use of evidence and that he was an advocate for his own theory.

The main arguments against drift were twofold: 1) geophysical, involving the rheological behavior of the continental crust versus the oceanic crust and the absence of a driving force; and 2) geological, paleontological, stratigraphic, and tectonic. Wegener put himself in a bind by proposing that the continents (sial) plowed through the ocean realm (sima), while at the same time the resistance of the sima caused fold and thrust belts in the front of the solid continents by resistance to the sima – the westward drift of the Americas and their western mountain ranges being the main examples. Wegener countered this problem as follows in later editions of his book: "The solution of this apparent contradiction lies in the great dimensions of the Earth and in the long periods of time."[4] Citing Maxwell on viscous fluids, he noted that a substance (for example sealing wax) may behave as a solid under a rapid impulse (e.g., an earthquake), but under a slow impulse it behaves as a viscous fluid. Nevertheless, under a slow impulse, a viscous sima still cannot crumple the front of an advancing solid continental block and at the same time flow. The fact that he ignored the oceanic crust as consisting of solid basalt made matters worse.

A second objection was that mountain-building was episodic in time, while the drifting of the continents does not produce episodic events. In a similar vein, if part of Gondwanaland broke up in the Cretaceous, why is this period tectonically quiet, with no mountain-building events? Wegener's drift hypothesis implies mountain-building events should occur continuously during drift, but this is not observed.

The physicist and geologist John Joly of Trinity College, Dublin, who we met in Chapter 2 for his work on the chemical age of the oceans, gained much support among the participants of the meeting. He suggested a modified version of Wegener's hypothesis involving episodic melting and episodic Continental Drift due to heating by radioactivity. In this hypothesis, continents acted as a thermal blanket over the sima causing melting and topographic highs. The continents then slid down off the topographic highs allowing the heat to escape by conduction causing cooling, and so on cyclically, producing episodic tectonic events. Many of the participants thought this idea deserved further work.

The objection by Bailey Willis was that the eastern side of westward drifting continents should also show tensional features, just as the western side shows compressional features. This argument can be overcome, however, if convection of the underlying sima is invoked so that the continents are floating on a convecting substratum, as Holmes had suggested in his 1929 paper.

Another major objection was that Continental Drift had no viable driving force. Wegener attributed the westward drift to tidal forces of the sun and moon, and the equatorial drift as due to the Earth's equatorial bulge. Others have pointed out that the emplacement of Alpine nappes and repeated glaciations were two examples of phenomena that were accepted at the time without known mechanisms or causes, so that Continental Drift could be recognized as valid without a causal mechanism; du Toit was of this opinion.

Geological and paleontological objections were many.[22] If the Atlantic opened in the Quaternary glacial period, where were the young fold belts of this age? Chamberlain was blunt on this point, saying, "The matching of the glacial moraines is ludicrous." Wegener did unfortunately match European and American glacial moraines in his first chapter, implying Europe and North America were recently connected. Similarly for the drifting away of Australia: Where were the young fold belts in Australia? More than one commentator suggested that a 50 percent to 75 percent similarity of species on the various continents would be expected if they formed a single continent, but only 5% showed such similarity. They suggested land bridges were a better explanation (even though isostasy ruled out sinking land bridges).

Charles Schuchert used plasticine outlines of the continents on a globe to show that if North America and Europe were juxtaposed that a large gap opens up between Alaska and Siberia, which had been joined since Cambrian time. Wegener replied that if North America was rotated rather than translated, no gap appears. Schuchert also noted a poor correlation of the geology between Ireland and Newfoundland that Wegener had placed together. Today that geology is regarded as consistent with the Wegener fit.

These are just a sample of the objections that came to light at the American Petroleum geologists' symposium of 1928. The debate was interrupted by World War II, but continued until 1944 when the *American Journal of Science* published several discussions against

Wegener's Hypothesis.[24, 25] One of these contributors was by Bailey Willis with a discussion entitled: "Continental Drift, Ein Märchen" (A Fairytale).[24] Alexander du Toit answered these objections in the *American Journal of Science* in 1944.[26]

In Europe a similar but less damning discussion of drift played out in the pages of *Nature Magazine*,[27] the *Geological* Magazine,[28] and the *Geographical Journal*.[29] A meeting in 1923 of the British Association in Hull, England, discussed the Wegener hypothesis, which produced a lively but inconclusive discussion.[30] All participants, however, agreed that Wegener's north Atlantic was much too young.

PALEOMAGNETISM

William Gilbert, born in Colchester, England (1544–1603), was physician to Queen Elizabeth I., and of whom Galileo said "great to a degree that might be envied." He wrote *De Magnete* in 1600, the first scientific book on magnetism and possibly the first book in all of experimental science.[33] In it he recognized the Earth had a dipole field as if a giant bar magnet was embedded in it, aligned roughly with the Earth's spinning axis. He saw that the inclination (the angle from the horizontal) of a magnetic needle varied with latitude on the Earth, the magnetic field being horizontal at the equator and vertical at the poles. The Earth's magnetic field is now thought to be a dynamo that originates in the convection of the molten core.[34]

Paleomagnetism is the study of the Earth's magnetic field in the geological past and is undertaken by the study of the magnetization of rocks. In the simplest case, for example, when volcanic lava erupts at the surface and cools down below the Curie temperature (about 600°C), iron-rich minerals in the lava lock in the Earth's ambient magnetic field at the time of eruption. However, subsequent metamorphic or weathering events can superimpose new magnetic fields on the original field corresponding to younger times, leading to possibly erroneous interpretations. Secondary hematite (Fe_2O_3), for example, might overprint the primary field due to original magnetite (Fe_3O_4). Today there are laboratory techniques to detect these complications and remove these overprints.

Oriented rock samples of known age are collected in the field and the direction of the rocks' remnant magnetization is measured in the laboratory with a magnetometer. The latitude of the rock at the time of eruption can then be calculated (using the formula $\tan \theta = \frac{1}{2} \tan I$, where θ is latitude and I is inclination), and a magnetic pole can also be calculated which is close to the geographic pole. This is done for rocks of different age and the change in magnetic pole is plotted for different times. The resulting curve is called an apparent polar wandering path because it is not clear whether the continent to which the samples belong to actually moved or the magnetic poles moved; the magnetic data alone cannot distinguish between the two.

It was the French scientist P. Mercanton in 1926 who suggested that rock paleomagnetism could be used to test Continental Drift, but it was not until the 1950s that this was actually accomplished.[35] Keith Runcorn (1956) showed that the apparent polar wandering paths for Europe and North America were different so that the two continents must have moved relative to one another (Figure 6.4).[36] Shortly thereafter results from the southern hemisphere showed that Gondwanaland was united in the Carboniferous and its constituent parts fragmented in the Mesozoic. Studies of the Indian Deccan trap volcanics showed that India moved northward about 4,000–5,000 kilometers in early Tertiary (Eocene) time.[37] The authors of the latter study were, however, noncommittal in their review of the paleomagnetic results. They concluded the paleomagnetic results might be explained more plausibly by a rapidly changing magnetic field rather than by large-scale Continental Drift. These authors did, however, redeem themselves of their faulty conclusion later on when they constructed a paleomagnetic timescale (see Chapter 7).[38]

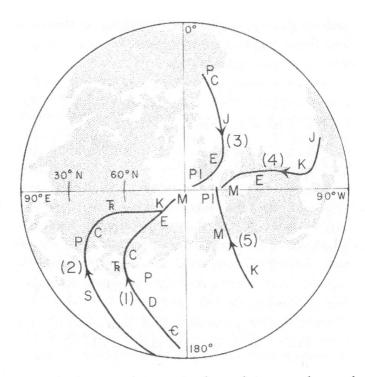

FIGURE 6.4 Apparent paleomagnetic polar wandering curves for several
different continents indicate major displacement of the continents
relative to one another during the Paleozoic, Mesozoic, and Cenozoic.
1. Europe; 2. North America; 3. Australia; 4. India; 5. Japan. S: Silurian;
D: Devonian; C: Carboniferous; P: Permian; T: Triassic; J: Jurassic;
K: Cretaceous; E: Eocene; M: Miocene; Pl: Pliocene.
Source: Cox, A. and Doell, R. 1960. Review of Paleomagnetism. *Bulletin of the
Geological Society of America*, v. 71, 645–768. Courtesy Geological Society
of America.

A symposium organized on Continental Drift as late as October
1965 by the Royal Society shows that both in Europe, and more so in
North America, many geologists and geophysicists were still not
onboard with the drift concept.[39] Tectonics was still in crisis mode.
In Part II of this book, we will see that paleomagnetism of a different
sort would lead to a new revolution in the study of Earth sciences.
Continental Drift had its day, and it would soon be replaced by New
Global Tectonics.

REFERENCES

1. Stow, G. W. 1871. On some points in South-African geology. *Quarterly Journal of Geological Society of London*, v. 27, 497–548.

2. Blanford, H. F. 1875. On the age and correlations of the plant-bearing series of India, and the former existence of an Indo-oceanic continent. *Quarterly Journal of Geological Society of London*, v. 31, 519–542.

3. Taylor, F. B. 1910. Bearing of the Tertiary mountain belt on the origin of the Earth's plan. *Geological Society of America Bulletin*, v. 21, 179–226.

4. Wegener, A. 1924. *The Origin of the Continents and Oceans*. Methuen, London. Trans., J. G. A. Skerl. First published in German 1915.

5. Frankel, H. F. 2012. *The Continental Drift Controversy:* Wegener and the early debate. v. 1, Cambridge University Press, Cambridge.

6. Greene, M. T. 2008. Alfred Wegener. *Complete Dictionary of Scientific Biography*, v. 25, 245–248. Charles Scribner's Sons, Detroit.

7. Jeffreys, H. 1929. *The Earth: It's Origin, History and Physical Constitution*, 2nd ed. Cambridge University Press, Cambridge.

8. Oreskes, N. 1999. *The Rejection of Continental Drift: Theory and Method in American Earth Science*. Oxford University Press, New York.

9. Chamberlin, T. C. 1890. The method of multiple working hypotheses. *Science*, v. 15, 92–96.

10. Dunham, K. 2008. Arthur Holmes. *Complete Dictionary of Scientific Biography*. v. 6, 474–476. Scribner's Sons, Detroit.

11. Holmes, A. 1913. *Age of the Earth*. Harper & Brothers, London.

12. Holmes, A. 1945. *Principles of Physical Geology*. Ronald Press, New York.

13. Holmes, A. 1929. Radioactivity and Earth movements. *Transactions of Geological Society of Glasgow*, v. 18, 559–606.

14. Wegmann, C. E. 2008. Émile Argand. *Complete Dictionary of Scientific Biography*, v. 1, 235 – 237. Charles Scribner's Sons, Detroit.

15. Argand, É. 1916. Sur l'arc des Alpes Occidentales. *Eclogae Geol. Helvetiae*, v. 14, 145–190.

16. Argand, É. 1922. La tectonique de l'Asie. Conférence faite á Bruxelles, le 10 août. *Congres géologique international, Belgique*.

17. Carozzi, A. V. 1977. *Tectonics of Asia*. Hafner Press, New York.

18. Wilson, J. T. 2008. Alexander du Toit. A Dictionary of Scientific Biography. *Charles Scribner's Sons*, v. 4, 261–263.

19. Du Toit, A. 1927. *A Geological Comparison of South America and South Africa*. Carnegie Institution, Washington D.C.

20. Du Toit, A. 1937. *Our Wandering Continents: An Hypothesis of Continental Drifting*. Oliver & Boyd, London.

21. Daly, R. A. 1926. *Our Mobile Earth.* Charles Scribner's Sons, New York.

22. Van Waterschoot van der Gracht, W. (ed.) 1928. *Theory of Continental Drift: A Symposium.* Chicago University Press, Chicago.

23. Gregory, J. W. 1929. *The Structure of Asia.* Methuen, London.

24. Willis, B. 1944. Discussion: Continental Drift, Ein Märchen. *American Journal of Science,* v. 242, 509–513.

25. Longwell, C. H. 1944. Further discussion on Continental Drift. *American Journal of Science,* v. 242, 514–515.

26. Du Toit, A. 1944. Discussion: Further remarks on Continental Drift. *American Journal of Science,* v. 242, 404–408.

27. Gregory, J. W. 1925. Review of *The Origin of Continents and Oceans* by Alfred Wegener. *Nature,* v. 115, 255–257.

28. Lake, P. 1925. Review of *The Origin of Continents and Oceans* by Alfred Wegener. *Geological Magazine,* v. 63, 181–182.

29. Lake, P. 1923. Wegener's hypothesis of Continental Drift. *The Geographical Journal,* v. 61, 179–187.

30. Wright, W. B. 1923. The Wegener Hypothesis: Discussion at the British Association, Hull. *Nature,* v. 111, 30–31.

31. Haworth, R. J. 2008. Harold Jeffreys. *Complete Dictionary of Scientific biography.,* v. 22, 38–42. Charles Scribner's Sons, Detroit.

32. Jeffreys, H. 1976. *The Earth: Its Origin, History and Physical Constitution,* 6th ed. Cambridge University Press, Cambridge.

33. Gilbert, W. 1600. *De Magnete.* Trans., P. F. Mottelay, 1893. Dover Publications, New York.

34. Bullard, E. C. and Gellman, H. 1954. Homogeneous dynamos and geomagnetism. *Philosophical Transactions of the Royal Society of London, ser. A,* v. 247, 213–255.

35. Mercanton, P. 1926. *Aimatation des basalts Groenlandais.* Academie Scieances Comptes Rendus, Paris, 859–860.

36. Runcorn, S. K. 1956. Paleomagnetic comparisons between Europe and North America. *Proceedings of the Geological Association of Canada,* v. 8, 77–85.

37. Cox, A. and Doell, R. 1960. Review of Paleomagnetism. *Bulletin of the Geological Society of America,* v. 71, 645–768.

38. Frankel, H. R. 2012. *The Continental Drift Controversy: Paleomagnetism and Confirmation of Drift.* v. 2, Cambridge University Press, Cambridge.

39. Blackett, P. M. S., Bullard, E. and Runcorn, S. K. (eds.). 1965. Continental Drift: A symposium. *Philosophical Transactions of the Royal Society of London,* v. 258.

7 Plate Tectonics

I shall consider this to be an essay in geopoetry.

– H. Hess, 1962.[1]

INTRODUCTION

While patrolling the northwest Pacific during World War II, the commander of the U.S.S. *Cape Johnson*, Harry H. Hess (1906–1969) often left the echo transponder on continuously as he crossed the Pacific Ocean in random traverses. Normally the sonar was only used while leaving and entering port, but as a geologist Hess was interested in the topography of the deep ocean floor. He later published a paper in the *American Journal of Science* (1946) stating that during service in the Navy he had discovered over one hundred flat-topped volcanoes that stood thousands of meters above the sea floor; he called them *guyots*, named after a flat-topped building on his campus (which in turn was named after a Swiss geographer).[2] He inferred the flat tops were due to wave erosion when sea level was lower. As an example of how little was known about the ocean floor at the time, Hess concluded that the guyots were Precambrian in age because no reefs were present, in contrast to Darwin's younger atolls (Chapter 1). We now know that the oldest rocks in the Pacific Ocean basin are Jurassic in age, about 170 million years old, not Precambrian. Over a decade later, Hess proposed the idea of sea-floor spreading, which was subsequently confirmed by the Vine-Matthews hypothesis based on magnetic stripes at mid-ocean ridges.[1]

The discovery of plate tectonics involved three narratives that were going on at approximately the same time in the 1960s, namely: the study of the magnetic signature of the ocean crust, the study of deep-focus earthquakes around the northern Pacific rim, and recognition of transform faults in ocean basins. Only a relatively few institutions were involved in the fundamental discoveries of plate tectonics, and they included Cambridge University, England; Princeton University, New Jersey; the Lamont Geological Observatory (now Lamont–Doherty Observatory), Columbia University, New York; and Scripps Institution of Oceanography, California.[3] In 1965 the paths of several individuals involved in plate tectonics crossed at Cambridge. Harry Hess visited from Princeton, as did Tuzo Wilson from the University of Toronto. Research students at Cambridge who were to make important contributions included Fred Vine, Dan McKenzie, Robert Parker, and John Sclater. Drummond Matthews was already on the faculty (or staff, as they say in England) and was to become Vine's doctoral thesis advisor. Three seismologists at Lamont would later crystallize the plate tectonic synthesis in an amazingly short period of time.[3]

Earlier, Hugo Benioff of the California Institute of Technology worked on deep crustal earthquakes[4] in the late 1950s which would later lead to discovery of destructive plate boundaries (subduction zones) by two seismologists Jack Oliver and Bryan Isacks at Lamont in 1968.[5] Together with constructive oceanic plate boundaries (seafloor spreading ridges) recognized by Hess[2] and Dietz[6] in the early 1960s and by Vine and Matthews in 1963,[7] these discoveries led to the New Global Tectonics. Tuzo Wilson described a third type of plate boundary: transform faults.[8]

Two instrumental or technological advances greatly aided in the discovery of plate tectonics. The first was the fluxgate magnetometer invented in 1936, which when airborne was used to detect submarines during World War II. After the war the first oceanic magnetic anomalies were discovered using this type of magnetometer towed behind a ship.[7] The second important advance came with the installation of the World-Wide Standardized Seismograph Network which commenced operation in 1962.[3] This network included calibrated seismographs of both long and short periods with high sensitivity and global coverage. The network was originally set up to monitor nuclear explosions during the ban on the testing of nuclear weapons. This network greatly improved the ability of seismologists to determine the sense of shear and orientation of the fault that caused an earthquake. These so-called first-motion studies allowed the determination of the type of stress at the three principal types of plate boundaries and played a very important role in understanding global tectonics. Data provided by the Deep Sea Drilling Project (DSDP) also yielded important information to support plate tectonics.[9]

OCEAN FLOOR RIFTS

In the late 1920s three research vessels, the *Carnegie*, the *Meteor*, and the *Dana* (American, German, and Danish in origin, respectively), made discoveries of ocean ridges in three different oceans using echo soundings; these ridges represented submarine mountain ranges that were topographically several kilometers above the surrounding ocean floor and hundreds of kilometers in length, but they were thought to be isolated features. By the 1960s a substantial amount of new geophysical data had been collected in the ocean basins worldwide including gravity, seismic, and heat flow data. Bruce Heezen of Lamont published a paper for a general audience in *Scientific American* in 1960 describing a worldwide system of connected oceanic rifts 64,000 kilometers (40,000 miles) in length that showed high heat flow and shallow earthquake foci largely confined to the rift zone.[10] The rifts were now recognized to be a global feature and not isolated phenomena. Henry William Menard, in his book *The Ocean of Truth*, provides a detailed account of early research on ocean basin geology in which Menard himself was deeply involved.[11,12]

Because subduction zones had not yet been discovered and because all the rifts were undergoing extension, Heezen suggested the Earth must be expanding. He attributed the cause of the high heat flow and topography at the ridges to the upwelling limbs of convection currents in the mantle. Robert Dietz in 1961 also tied mantle convection to upwelling at ocean ridges and down welling at Pacific-type continental margins, without the necessity for earth expansion.[6]

In 1961 Henry Menard of the Scripps Institute, California, described in substantial detail the high heat flow and shallow seismicity of the East Pacific Rise (rise and ridge are here interchangeable terms) and noted the low heat flow of deep ocean trenches supporting the idea that these were the down going limbs of the convection cells.[12] Menard, in his book, also discusses those who supported expansion of the Earth over geological time as an explanation for the present-day distribution of the continents.[11]

Harry Hess in his 1962 paper did not like the earth expansion idea, and agreed that the down going limbs of the convection currents took care of the material balance produced at the expanding ridges including their sedimentary cover that was eventually accreted to the continents.[1] Hess probably overemphasized the role of hydration of mantle peridotite to produce serpentine at the axes of the ridges (his doctoral thesis was on the hydration of ultramafic rocks). He recognized that sea-floor spreading (the term was first used by Dietz) "wiped the slate clean" on a relatively short geological time scale which explained the absence of Paleozoic-aged ocean ridges and of old oceanic sediments.[1] Who took priority over the sea-floor spreading concept is discussed by Menard in his book, and he favors Hess; Hess had earlier circulated a preprint of his manuscript in 1960, but Dietz had no recollection of this or of discussing the idea with Hess.[11]

In the 1950s several paleomagnetic studies showed magnetic polar reversals in volcanic rocks where the magnetic field direction appeared to be flipped by 180 degrees so that magnetic north became magnetic south; one of these early studies was on basalt flows in Iceland by a student at Cambridge.[13] Initially these reversals were interpreted as due to the instability of the magnetic signature in individual rock samples (called self reversals), but when enough data was collected, the reversals were seen to be global in extent and also contemporaneous. These data were attributed to reversal of the Earth's entire magnetic field, even though a good explanation for the reversals did not yet exist.[3] Workers at the United States Geologic Survey identified four major reversals and about a dozen or so short reversals over a period of 4.5 million years. This allowed the youngest part of the ocean floor close to the spreading center to be dated.[14] One of these reversals occurred during the Pleistocene and is named the Matuyama anomaly after the Japanese scientist who showed that a large number of Pleistocene-aged volcanic rocks from Japan indicate reverse polarity. This reversal ended about 0.8 million years ago, and since then the Earth has been under normal polarity, called the Brunhes period, named after the French physicist. The magnetic reversals were dated using the potassium-argon (K-Ar) radiometric decay system (see Chapter 8) on volcanic lava flows from around the world. Gradually, the age of the reversals was established for oceanic rocks as far back as 160 million years ago, corresponding to some of the oldest rocks in the oceans. The recognition of global magnetic polar reversals led the way to explain magnetic anomalies at mid-ocean ridges, thereby confirming the reality of sea-floor spreading as suggested by Dietz (1961) and Hess (1962).

Two English scientists working at Cambridge University, Fred Vine and Drummond Matthews, interpreted positive and negative magnetic anomalies centered on mid-ocean ridges that produced linear magnetic stripes parallel to the ridges as due to the magnetic reversals just described above.[7] Some of the data they used was produced aboard the H. M. S. *Owen* in 1962 by criss-crossing the northern Indian Ocean at the Carlsberg Ridge, towing a magnetometer behind the ship; they also used similar data from the Mid-Atlantic Ridge. According to the sea-floor spreading hypothesis, new basalt was extruded at the seafloor at mid-ocean ridges due to the upwelling of convection currents in the mantle. As the basalts cooled below their Curie temperature (~600°C), they locked in the ambient magnetic field of the Earth (whether normal or reversed) and gradually moved away from the rift zone at the rate of centimeters per year to be replaced by new basalt at the ridge axis. If the Earth's magnetic polarity changed, these new basalts would record the new magnetic field producing the reversed and normal anomalies observed in stripes parallel to the ridge. As Fred Vine noted, the oceanic crust acted as a tape recorder of the Earth's magnetic history. This eventually would allow a magnetic stratigraphy to be constructed for the entire history of the ocean basins. That the sediments deposited in the ocean basins should become older toward their margins is a corollary of the Vine-Matthews hypothesis, which was later confirmed by deep-sea drilling.

The Deep Sea Drilling Project (DSDP) began in 1968 with the launch of the *Glomar Challenger*, a drilling ship outfitted to obtain cores of sediment from the deep oceans; it also had a satellite global positioning system, novel for that time. Funded by the United States National Science Foundation, it was a highly successful scientific project over a fifteen-year period.[9] The *Glomar Challenger* logged 600,000 kilometers (375,000 miles) and sampled 19,000 cores from 624 sites. One of its most successful projects was Leg 3, which sampled seventeen cores in the south Atlantic Ocean from Rio de Janeiro to the Mid-Atlantic Ridge at about 30°S latitude. The results of the study showed that the sediments became older from zero at the ridge to 75 million years old with distance from the axis of the ridge (indicating a half spreading rate of 2 cm/yr), strongly supporting the Vine-Matthews hypothesis (Figure 7.1).

The near perfect symmetry of the magnetic anomalies detected immediately south of Iceland on the Reykjanes Ridge, which is reproduced in many introductory geology textbooks in the context of the Vine-Matthews hypothesis, was not published until 1966, three years after the Vine-Matthews paper appeared, so these authors were unlikely to have been aware of the Icelandic data (and it should be noted that they don't mention these data in their paper). The Icelandic aeromagnetic data were collected by a magnetometer suspended from a plane and were much more accurate than the data collected at sea by Vine and Matthews in the Indian Ocean. The Icelandic data therefore do not appear to have contributed to the Vine-Matthews hypothesis, but those data supplied strong additional confirmation afterward.

In 1963 a Canadian geophysicist (Lawerenc W. Morley) was working with previously collected aeromagnetic anomalies in the

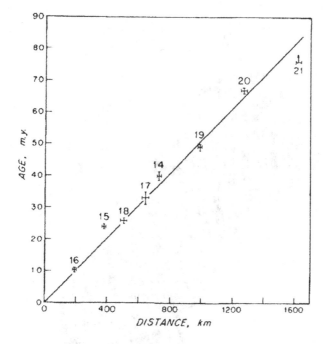

FIGURE 7.1 Leg 3 of the Deep Sea Drilling Project in the south Atlantic (about 30°S latitude) showed a linear relationship between distance from the Mid-Atlantic Ridge and the age of sediments immediately above basement. These data strongly supported the Vine-Matthews hypothesis. *Source:* Maxwell, A. E. et al. 1970. *Initial Reports of the Deep Sea Drilling Project,* v. 3. U.S. Government Printing Office, Washington D.C.

eastern Pacific Ocean offshore Vancouver Island that showed similar zebra stripes to those offshore Iceland, but were much more complex and less symmetric (Figure 7.2).[15] Morley had essentially the same idea as Vine and Matthews in interpreting these anomalies. Unfortunately for him, his paper was rejected by two journals (*Nature* and the *Journal of Geophysical Research*) in 1963 for being too speculative. One possible reason for Morley's rejection is that he was not using newly acquired data (the data he used was collected earlier by others),[15] whereas Vine and Matthews were using newly collected data. In addition, the anomalies were not symmetrical about a single spreading axis. Some writers now refer to this hypothesis as the

FIGURE 7.2 Aeromagnetic survey offshore Vancouver Island in the eastern Pacific ocean. The positive anomalies are black and negative anomalies white. The anomalies are offset by transform faults showing a complex pattern.

Source: Raff, A. D. and Mason, R. G. 1961. Magnetic survey off the west coast of North America, 40° N latitude to 50° N latitude. *Geol. Soc. Am. Bull.*, v. 72, 1267–1270. Courtesy Geological Society of America.

Vine-Matthews-Morley hypothesis. Morley's paper is discussed in detail by Frankel,[16] and is partly reproduced in Cox[3].

TRANSFORM FAULTS

Henry Menard showed the existence of east-west trending fracture zones that offset the East Pacific Rise (or Ridge) by several hundred kilometers (such as the Mendocino, Galapagos, and Easter fracture zones).[11] These offsets of mid-ocean ridges are global in extent and were later explained as a new type of plate boundary that Tuzo Wilson called transform faults.[8] They differ from transcurrent (or strike-slip) faults in fundamental ways. In oceanic transform faults, the sense of motion along the fault is (confusingly) opposite to the offset of the ridge itself because new crust is being created at the ridge axis (Figure 7.3A). Several types of transform faults exist: they can connect offsets of ocean ridges, such as the mid-Atlantic fracture zones (ridge-ridge type), or they may connect ocean ridges to island arcs (ridge-arc type). Wilson showed that six types of transform faults are possible (twelve if you reverse the sense of motion). He interpreted the San Andreas fault as a transform fault connecting the East Pacific Ridge in offshore southern California to the Juan de Fuca Ridge in offshore northern California.

Seismic studies undertaken by Lynn Sykes at Lamont at about the same time (1963) showed that the seismic activity on transform faults was confined to the segment of the transform fault *between* the ridges – the fault segments outside the ridges were aseismic, supporting Wilson's interpretation (Figure 7.3B).[17] Furthermore, using first motion or focal mechanism seismic studies, Sykes subsequently showed that the motion on the faults was as Wilson described, namely opposite to the offset of the ridge.[18] These seismic studies confirmed Wilson's interpretation of transform faults, leaving few doubters as to the reality of these counter-intuitive faults. Transform faults will continue to confound introductory geology students for generations to come. The understanding of the geometry of transform faults would later provide the key to understanding the relative motion of the tectonic plates.

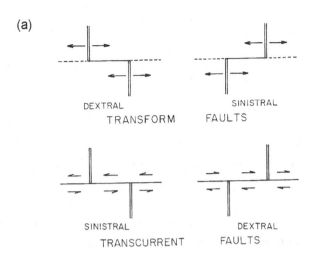

(a)

DEXTRAL
TRANSFORM   FAULTS   SINISTRAL

SINISTRAL
TRANSCURRENT   FAULTS   DEXTRAL

(b)

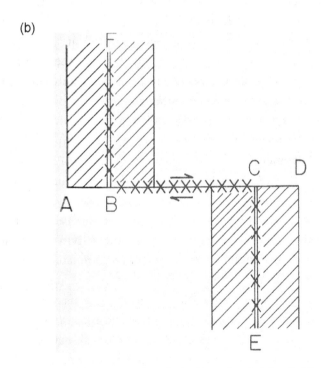

FIGURE 7.3 (a) Top: Transform faults (dextral and sinistral). The sense of motion on the fault is opposite to the offset of the ridge axis. Bottom: Transcurrent faults (sinistral and dextral). The sense of motion on the faults agree with the ridge offset.
Source: Sykes, L. R.1963. Seismicity of the South Pacific Ocean. *Journal Geophysical Research*, v. 68, 5999–6006. Courtesy Wiley.
(b) Seismic activity (X) on a spreading center is confined to the ridge axis and the transform segment between the offset ridge axis (BC). Fault segments beyond the ridge axis (AB; CD) are aseismic. These relationships confirm Tuzo Wilson's predictions for transform faults.
Source: Sykes, L. R.1967. Mechanisms of earthquakes and nature of faulting on mid-ocean ridges. *Journal of Geophysical Research*, v. 72, 2131–2153. Courtesy Wiley.

SUBDUCTION ZONES

Arthur Holmes in his 1929 paper (Chapter 6) on convection discusses down going convection currents: "Evidence of foundering blocks may be forthcoming from the occurrence of deep earthquakes (100 kilometers or more) off the coast of Japan." It was indeed deep earthquakes that established the existence of subduction zones where oceanic crust and the upper mantle (which together are referred to as oceanic lithosphere) descended into the deep mantle. Hugo Benioff at the California Institute of Technology in 1954 summarized the existing data on deep earthquakes by plotting their horizontal distance from volcanic arcs and deep ocean trenches (Figure 7.4).[4] Although he did not explicitly associate these deep earthquakes with down-going slabs, it would not be long before geophysicists associated these earthquakes with destructive plate margins (now called Benioff zones), complementary to the constructive ocean ridge plate boundaries described earlier.

The concept of tectonic plates had not yet emerged in 1954. The first mention of rigid tectonic plates, including ocean ridges and island arcs, appears to be Tuzo Wilson's paper in 1965 on transform faults already referred to; in that paper, arrows show the direction of motion of the plates. Seismometers installed in the Tonga trench showed that deep-focus earthquakes occurred on the upper surface of the descending slab (about 100 kilometers thick) within the Fiji region of the Pacific Ocean by Lamont researcher Bryan Isacks, which allowed the first lithospheric subduction zone to be delineated by Oliver and Isacks.[5] These authors suspected a similar pattern of seismicity beneath other deep-sea trenches for which data existed. Enough was now known that a new global synthesis could be attempted.

THE NEW GLOBAL TECTONICS

Two important papers entitled *Seismology and the New Global Tectonics*[19] and *Sea-floor Spreading and Continental Drift*[20] were published in 1968 by researchers all working at the Lamont

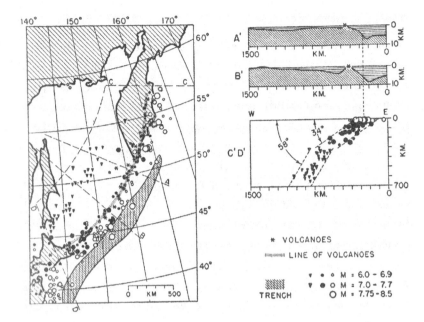

FIGURE 7.4 Left: Map of the Kurile-Kamchatka region that connects northern Japan and the Kamchatka Peninsula, showing epicenters of shallow (circles), intermediate (dots), and deep-focus earthquakes (triangles). The deep-sea trench is shown (crosshatch). Right: Composite profile of earthquake foci to 700 kilometers depth. The zone of deep and intermediate earthquakes dips toward the continent at two different angles. The stars represent active volcanoes.

Source: Benioff, H. 1954. Orogensis and deep crustal structure: additional evidence from seismology. *Geological Society of America*, v. 65, 385–400. Courtesy Geological Society of America.

Observatory. The latter paper by the French scientist Xavier Le Pichon identified six large rigid "blocks" (now called tectonic plates), namely: India, Antarctica, Africa, Pacific, America, Eurasia. Using the geometric principles for tectonics on a sphere and ocean floor magnetic anomalies, Le Pichon showed how the six rigid plates moved relative to one another together with their velocities on the globe. He used numerical methods on a digital computer with various types of Mercator projections for a visual display of the different plates. The geometrical principles he used were developed earlier by Jason Morgan at Princeton University[21] and Dan McKenzie and Robert Parker at the Scripps Research Institute,[22] and are briefly outlined here.

The relative motion between two plates on a sphere can be described by an axis of rotation, a line through the center of the Earth that can be simply specified by its latitude and longitude at the point it pierces the surface; this is referred to as Euler's theorem, after the eighteenth-century Swiss mathematician. It was Cambridge University scientists who first applied Euler's theorem in their fit of Africa and South America.[23] If three plates meet at a point, then the relative motion between two plates determines the relative motion of the third. McKenzie and Parker used earthquake first motion studies around the Pacific to find the pole of rotation for the Pacific plate relative to Asia.[22] Jason Morgan at Princeton developed the method of using transform faults to determine poles of rotation.[21] He saw that transform faults lie along small circles (those not passing through the center of the Earth), and that great circles (those that do pass through the center of the Earth) intersect these small circles at a right angle at the pole of rotation of two plates. This is the most accurate way to determine poles of rotation. Robert Parker also showed that the Mercator map projection was especially useful in determining the axis of rotation between two plates. As geography students know, the Mercator projection is a cylindrical projection and is normally thought of as a projection onto a vertical cylinder touching the Earth at the equator. But the cylinder can have any orientation, and Robert Parker had written a general computer program to do just that. He then recognized that if the pole of the Mercator projection was the same as the pole of rotation between two plates, then transform faults became horizontal lines on the projection – if the pole of rotation was correct.[22] Le Pichon used these geometric principles to solve the motion of his six plates on a sphere together with ocean floor magnetic anomalies.[20] His results agreed with those based on seismic first motion studies in the paper by Isacks, Oliver, and Sykes.[19] It was only 1968, and the basic geometry of plate tectonics was already in place.

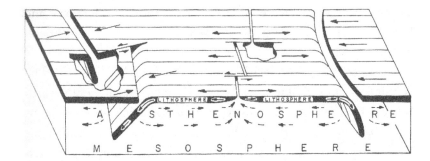

FIGURE 7.5 The now iconic cartoon figure from the paper entitled
"Seismology and the New Global Tectonics" by Isacks, Oliver, and Sykes.
It is notable for the absence of continents. The pattern of circulation in
the asthenosphere is also unusual because it is opposite to the motion
of the overlying plates.
Source: Reference 19: Isacks, Oliver and Sykes, 1968.

Perhaps the most iconic diagram of New Global Tectonics is a
figure by Isacks and colleagues reproduced here in Figure 7.5. It is
noteworthy for two reasons: First is the absence of continents, which
underscores the point that plate tectonics was essentially a theory
developed solely on evidence from the ocean basins but d later proved
powerful in also explaining continental tectonics. The second unusual
feature of the figure is that the motion in the asthenosphere (the weak
layer beneath the lithosphere) does not conform to normal thermal
convection – in fact, the motion in the asthenosphere is *opposite* to
that in the overlying plates – and there is no return circulation pattern
in the asthenosphere; the arrows point in the same direction at both
deep and shallow levels. The authors suggest that the plates themselves may determine the circulation in the asthenosphere. They
presented data suggesting a relationship between the velocity of the
plates and the length of the attached subduction zone. Subsequent
workers examined the gravitational forces that act upon the plates
and did indeed show that subduction zones acted to pull the plate into
the mantle (slab pull), and also that the topographic highs of the ridges
acted to push the plates away from the ridge (ridge push).[24,25]

Isacks, Oliver, and Sykes also showed that the Earth's plate boundaries are largely defined by the distribution of both shallow-focus earthquakes (at ocean ridges) and deep-focus earthquakes (at subduction zones). They also show that the stress within down-going slabs is compressive parallel to the slab, an issue which was in dispute at the time. Both Bryan Isacks and Lynn Sykes were graduate students of Jack Oliver at Lamont where most students and faculty at the time were "fixist" and did not take Continental Drift seriously. The historical details of this remarkably successful small research group and their conversion to mobilism is recounted by Henry R. Frankel.[16]

As noted, the majority of Earth scientists (geologists, geophysicists, and oceanographers) readily accepted plate tectonics as a valid theory, with scientists in the former Soviet Union being the notable exception.[26] The influential Cambridge geophysicist Harold Jeffreys was also a notable exception. In the sixth edition of *The Earth*, published in 1976, he did not accept the Vine-Matthews sea-floor spreading hypothesis. He also maintained that the lithosphere was too strong to bend into subduction zones.[27]

The American Association of Petroleum Geologists (AAPG) was interested in Wegener's drift hypothesis (Chapter 6), and, not surprisingly, they were also interested in the new plate tectonics theory and published a memoir in 1974 entitled *Plate Tectonics: Assessments and Reassessments*, which grew out of a symposium held in 1971.[28] A large number of contributing papers expressed doubts about the theory, including the Russian scientist Vladimir Beloussov, Harold Jeffreys himself, and several papers by Arthur Meyerhoff and Howard Meyerhoff and one by J. C. Maxwell. Some of the doubts expressed by these authors are briefly outlined here.

Arthur Meyerhoff and Howard Meyerhoff rejected the young age of the ocean floor magnetic anomalies.[29] They plotted the locations of known oceanic magnetic anomalies and noted that over half of them were concentric to ancient cratons, claiming that they intersect the continents and so are Precambrian in age. They went on to list radiometric age determinations of rocks from the ocean basins. Out of approximately 140 radiometric dates, all are Cenozoic in age with only four aberrant dates ages giving Precambrian ages. They reiterated their conclusion that the ocean basins are Precambrian in age, ignoring the far more numerous Cenozoic dates. They further examined the results of Leg 3 of the Deep Sea Drilling Project shown in Figure 7.1. Because the drill cores stopped at the sediment-basement interface, these authors maintain the linear relationship between distance from the ridge and sediment age "prove nothing."[29] It would be easy to dismiss these authors as fringe cranks, but Jeffreys in *The Earth* (sixth edition) quotes extensively from these authors' papers in his own rejection of plate tectonics, indicating they had substantial influence at least on Jeffreys, who was more of a mathematician than a geologist.

More plausible objections were made by J. C. Maxwell.[30] He noted the absence of strong deformations in trench sediments and that normal faults predominated rather than thrusts, as would be expected. Isacks et al. however had already explained the normal faulting was due to bending of the down-going plate causing local extension (normal faulting) in the sediments. Maxwell also noted the East Pacific Rise magnetic anomalies were not symmetric about a single spreading center. But Vine (1966) had shown earlier that there were three short offset ridges associated with the East Pacific Rise (Figure 7.2).[31] Beloussov questioned the width of magnetic anomalies as being inconsistent with the then known ages of recent magnetic reversals.[32] But it appears he was using the earlier magnetic stratigraphy that was preliminary and later improved by discovery of additional short reversals which eliminated the inconsistencies.[13] Beloussov published a more detailed rejection of sea-floor spreading in a 1970 paper in *Tectonophysics*, which would have had a wider readership compared to the AAPG memoir.[33]

IMPLICATIONS FOR MOUNTAIN BUILDING

A paper entitled *Mountain Belts and the New Global Tectonics* by John Dewey (Cambridge University) and John Bird (New York

University at Albany) appeared in 1970 only two years after the seminal paper by Isacks et al., so that geologists lost no time in applying plate tectonic principles to the evolution of the continents.[34] Finally, a viable theory of mountain-building had arrived and the long-standing tectonic crisis outlined in Chapter 5 came to an end. The 1970 paper is notable for its numerous excellent illustrations and the fact that it conceptually makes several breakthroughs in geology. The paper focused on continent-ocean boundaries, as this is the locus of mountain-building either due to continent-continent collision or continent-island arc collision as a result of subduction.

The long standing problem of geosynclines, the long linear belts of exceptionally thick sediments, first identified in the Appalachians by James Hall (Chapter 5), and how and why they become the locus of intense folding and igneous activity was solved by the Dewey and Bird paper. The authors first show the sediments of the Appalachian mio-geosyncline and eugeosyncline of Marshall Kay (see Figure 5.5) are very similar to the inner and outer margins of the present-day passive Atlantic margin. This passive margin is transformed into a subduction zone as the Iapetus (Ordovician aged) oceanic crust gets older and gains in density and sinks into the mantle. This leads to shortening of the passive-margin sediments and melting of the down-going slab, focusing igneous activity on the former geosyncline. The result is a mountain belt marginal to the continent, such as the Appalachians where the geosynclinal sediments become accreted to the continental margin and intruded by island arc volcanics. Readers familiar with northern Appalachian geology will recognize the Bronson Hill Anti-clinorium in New England as an Ordovician island arc that formed over a closing Iapetus Ocean. The role of subduction at continental margins in producing Cordillera-type orogenic belts (e.g., the Andes) is also addressed in the Dewey and Bird paper, particularly with regard to emplacement of thrust sheets onto the continental interior, in which basement may or may not be involved. Continent-continent collision is also discussed in the context of the only modern example: the India-Asia collision. Plate tectonic models of all the major mountain belts quickly appeared in the geologic literature, including models of orogenic belts going back into early earth history.[35]

IMPLICATIONS FOR IGNEOUS ACTIVITY AND TECTONIC SETTING

Plate tectonics also revolutionized our understanding of the evolution of igneous rocks. This occurred mainly in three different tectonic settings, with increasing complexity in terms of petrogenesis, namely mid-ocean ridges, ocean-ocean subduction zones, and ocean-continent subduction zones. Sampling of the ocean floor showed that the dominant igneous rock there is tholeiite basalt, also known as mid-ocean ridge basalt (or MORB), and is remarkably uniform in composition and consists largely of pyroxene and calcic feldspar (in the ratio of 60–40 percent) with occasional olivine. This led to a simple two-stage model for the origin of MORB: partial melting of the upper mantle, leading to a magma chamber at the mid-ocean ridge, followed by fractional crystallization of pyroxene and feldspar during cooling leading to the MORB composition.[36] Some of the best evidence for the structure of the oceanic crust comes from the study on land of mafic and ultramafic rocks (called ophiolite suites), seen in several mountain belts including the Alps (e.g., Troodos in Cyprus) and the Appalachians of western Newfoundland (e.g., Bay of Islands), where they are interpreted to represent tectonic slices of oceanic crust that was obducted (as opposed to subducted) onto the continental margin.[37] These rock suites commonly consist, from bottom to top, of ultramafic rocks overlain by gabbro injected by dikes, further overlain by pillow basalts and sediments that correspond to the upper mantle and seismic layers 3, 2, and 1, respectively identified at mid-ocean ridges (Figure 7.6). That the oldest ophiolites known are about two billion years old suggests plate tectonics has been operative since that time.[38]

The igneous rocks in oceanic island arc settings are more complex and range in composition, with increasing silica (SiO_2), from basalt to andesite to rhyolite – referred to as the calc-alkaline trend. In the case of several Pacific island arcs, the volcanic rocks become

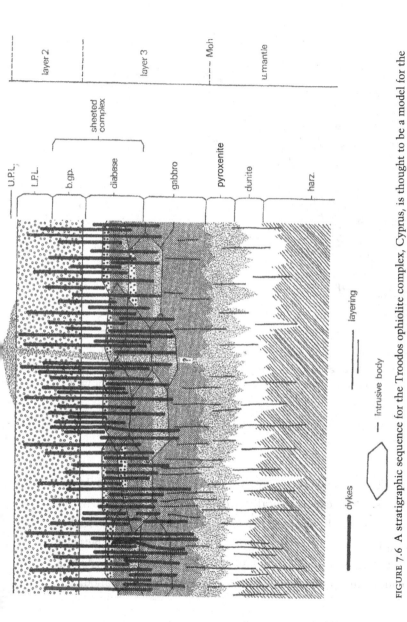

FIGURE 7.6 A stratigraphic sequence for the Troodos ophiolite complex, Cyprus, is thought to be a model for the structure of a mid-ocean ridge spreading center. Column at right indicates layers determined by seismic studies of mid-ocean ridges. UPL: upper pillow lavas; LPL: lower pillow lavas; BGP: basal group; Harz: harzburgite (olivine-pyroxene upper mantle rock); Moh: Mohorovicic boundary (crust-mantle interface).

Source: Moores, E. M. and Vine, F. J. 1971. The Troodos Massif, Cyprus and other ophiolites as oceanic crust: Evaluation and implications. *Philosophical Transactions of Royal Society, A,* v. 268, 443–467. Royal Society London with permission.

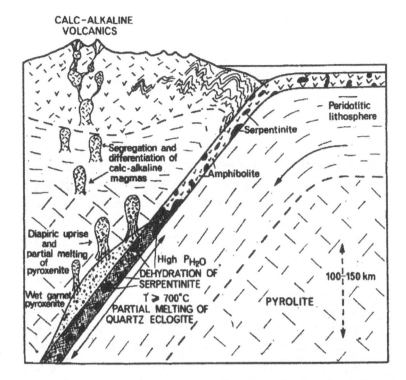

FIGURE 7.7 Igneous activity at a destructive plate boundary indicates partial melting of a down-going slab (after conversion of basalt to eclogite) produces calc-alkaline magmas that are accreted to the overlying plate to form a volcanic island arc.

Source: Ringwood, A. E. 1974. The petrological evolution of island arc systems. *Journal of Geological Society of London.*, v. 130, 183–204. Courtesy Geological Society London.

more potash (K_2O) and silica rich with increasing depth to the Benioff zone.[39] Australian experimental petrologists have also shown that partial melting of the down-going slab, after basalt is converted to eclogite (pyroxene and garnet-bearing rock) at a depth of between 100 kilometers and 300 kilometers, yields a calc-alkaline magma, which can in turn produce sialic crust such as diorite or granodiorite (Figure 7.7).[40]

A more complex tectonic setting is that of subduction at a continental-ocean plate boundary, where the additional variable is the potential involvement of the continental crust itself in the melting process together with any sediments involved in subduction. The question of the origin of granite batholiths is central to this problem, and was briefly addressed in Chapter 4. The origin of the granatoid rocks of the Sierra Nevada batholith of the western United States (most of which are diorite and granodiorite, as opposed to granite), and those of the Andes, for example, are still a significant scientific problem for geologists despite the tectonic framework provided by plate tectonics. The experimental work of Tuttle and Bowen outlined in Chapter 4 showed that in the presence of water, granitic melts can be produced at temperatures as low as 650°C, so that some granites can be produced by melting of sediments in the continental crust. Two Australian geologists (Bruce Chappell and A. J. R. White) in 1974, working in the Tasman orogenic belt of eastern Australia, proposed a classification of granitic rocks into two types: S-type (sedimentary source) and I-type (igneous source).[41] The S-type is mainly granite in composition and has accessory minerals rich in aluminum, such as micas and garnet, and is also rich in silica. The I-type has a wider range in composition from felsic to mafic (e.g., granite to diorite), and often contains the mineral hornblende. The two types also have distinct strontium isotopic compositions, reflecting different sources (see Chapter 8). The I-type granitic rocks could be produced by the subduction processes outlined in this chapter at a continental margin or island arc, and the S-type could be produced simply by metamorphism of a thick sedimentary pile. The English geologist Wallace S. Pitcher has discussed granite type and tectonic environment in considerable detail.[42]

CONCLUSIONS

Plate tectonics is a theory that crystallized over a very short period of time between 1961 and 1968, and gained widespread acceptance shortly thereafter. How could such a major scientific revolution have occurred over such a short period? Thomas Kuhn's book *The Structure of Scientific Revolutions* coincidentally appeared at about the same time the theory of plate tectonics was developing, and some have compared the plate tectonic revolution to a paradigm shift equivalent to that of the Copernican revolution in astronomy – although that is probably an overstatement.[43] But the acceptance of plate tectonics is an example of a paradigm change as described by Kuhn in his 1962 book (Chapter 12). It might be argued, however, that Wegener's continental drift hypothesis had been fomenting among the geological and geophysical community for forty years or longer (Wegner's ideas became widely known in several languages around 1925). The minds of Earth scientists may have been preconditioned by Wegener's radical ideas, so that when additional evidence became available in the form of sea-floor spreading and subduction zones, scientists were ready to get onboard. The problem of a mechanism for plate tectonics still existed (as it did for Wegener), but plate tectonics did not require plowing of continents through a strong sima, which was one of the main objections to Continental Drift. Plate tectonics was now acceptable to the majority of scientists, with or without a causal mechanism. Those who rejected plate tectonics were relatively insignificant in number, although some were prolific and famous. It may be suggested that in reality the acceptance by the majority of Earth scientists of large-scale continental mobility took at least forty years to ferment, requiring Wegner's ideas first, followed by continental paleomagnetic results, before plate tectonics could be accepted. The evidence presented by the proponents of plate tectonics was nevertheless overwhelming.

Dietz, in his important 1961 paper, stated: "The concept proposed here, which can be termed the spreading sea-floor theory, *is largely intuitive,* having been derived through an attempt to interpret sea-floor bathymetry"[6] (emphasis added). Hess, in his equally important 1962 paper, said: "I shall consider this paper to be an essay in geopoetry."[1] Apparently two of the most important papers leading to plate tectonics in the early 1960s are based on intuition and poetry. This is surely a long way from Chamberlin's multiple working hypothesis method that he encouraged all scientists to follow and which Menard said he tried to follow.[11] It is even farther away from Sedgwick's earlier inveighing against Lyell's uniformitarianism (quoted in Chapter 1). Wegener did not indulge in geopoetry or intuition, but rather marshalled a scientific hypothesis with solid evidence, yet he was still rejected. Clearly a major shift over time had occurred in how scientists were allowed to undertake science. World War I and World War II had intervened between Wegener's hypothesis and plate tectonics theories in the early 1960s. That little progress was made on tectonic issues, such as the origin of mountain belts, since the late eighteenth century may have convinced many that intuition and poetry were worth a try, since apparently Chamberlin's approach had not worked. A more prosaic explanation is that in the counter-culture environment of the 1960s, it was popular to espouse intuition and poetry as guiding forces – particularly in the context of Mother Earth.

REFERENCES

1. Hess, H. H. 1962. History of ocean basins. In *Petrologic Studies: A volume in honor of A. F. Buddington* (Engel, A. E., James, J., and Leonard, B. F., eds.). *Geological Society America*, 599–620.

2. Hess, H. H. 1946. Drowned ancient islands of the Pacific basin. *American Journal of Science*, v. 244, 772–791.

3. Cox, A. 1973. *Plate Tectonics and Geomagnetic Reversals*: Selected readings, Alan Cox (Ed.). Freeman, San Francisco.

4. Benioff, H. 1954. Orogensis and deep crustal structure: additional evidence from seismology. *Geological Society of America*, v. 65, 385–400.

5. Oliver, J. and Isacks, B. 1968. Structure and mobility of the crust and mantle in the vicinity of island arcs. *Canadian Journal of Earth Sciences*, v. 5, 985–991.

6. Dietz, R. S. 1961. Continent and ocean basin evolution by spreading of the sea floor. *Nature*, v. 190, 854–857.

7. Vine, F. J. and Matthews, D. H. 1963. Magnetic anomalies over oceanic ridges. *Nature*, v. 199, 947–949.

8. Wilson, J. T. 1965. A new class of faults and their bearing on continental drift. *Nature*, v. 207, 343–347.

9. Maxwell, A. E., Von Herzon, R. P., Andrews, J. E., Boyce, R. E., Milow, E. D., Hsu, K. J., Percival, S. F. and Saito, T. 1970. *Initial Reports of the Deep Sea Drilling Project*, v. 3. U.S. Government Printing Office, Washington, DC.

10. Heezen, B. C. 1960. *The Rift in the Ocean Floor.* Scientific American, October, 98–110.

11. Menard, H. W. 1986. *The Ocean of Truth: A personal history of global tectonics.* University of Princeton Press, Princeton, NJ.

12. Menard, H. W. 1965. The world-wide oceanic rise-ridge system. *Philosophical Transactions of Royal Society of London, Series A,* v. 258, 109–122.

13. Hospers, J. 1951. Remnant magnetism of rocks and the history of the magnetic field. *Nature,* v. 168, 1111–1112.

14. Cox, A., Dalrymple, G. B., and Doell, R. R. 1967. Reversals of the Earth's magnetic field. *Scientific American,* v. 216, 44–54.

15. Raff, A. D. and Mason, R. G. 1961. Magnetic survey off the west coast of North America, 40°N latitude to 50°N latitude. *Geological Society of America Bulletin,* v. 72, 1267–1270.

16. Frankel, H. R. 2012. *The Continental Drift Controversy: Evolution into Plate Tectonics,* v. 4., Cambridge University Press, Cambridge.

17. Sykes, L. R. 1963. Seismicity of the South Pacific Ocean. *Journal of Geophysical Research,* v. 68, 5999–6006.

18. Sykes, L. R. 1967. Mechanisms of earthquakes and nature of faulting on mid-ocean ridges. *Journal of Geophysical Research,* v. 72, 2131–2153.

19. Isacks, B., Oliver, J., and Sykes, L. R.1968. Seismology and the new global tectonics. *Journal of Geophysical Research,* v. 73, 5855–5899.

20. Le Pichon, X. 1968. Sea floor spreading and continental drift. *Journal of Geophysical Research,* v. 73, 3661–3697.

21. Morgan, W. J. 1968. Rises, trenches, and crustal blocks. *Journal of Geophysical Research,* v. 73, 1959–1982.

22. McKenzie, D. P. and Parker, R. L. 1967. The North Pacific: An example of tectonics on a sphere. *Nature,* v. 216, 1276–1280.

23. Bullard, E. C., Everett, J. E., and Smith, A. G. 1965. Fit of the continents around the Atlantic. A symposium on continental drift (Blackett, P. M. S., Bullard, E. C., and Runcorn, S. K. eds.). *Philosophical Transactions of Royal Society of London Series A,* v. 258, 41–75.

24. Forsyth, D. and Uyeda, S. 1975. On the relative importance of the driving forces of plate motion. *Geophysical Journal of Royal Astronomical Society,* v. 43, 163–200.

25. Chapple, W. M. and Tullis, T. E. 1977. Evaluation of the forces that drive the plates. *Journal of Geophysical Research,* v. 82, 1967–1984.

26. Khain, V. E. 1991. Mobilism and plate tectonics in the USSR. *Tectonophysics,* v. 199, 137–148.

27. Jeffreys, H. 1976. *The Earth: Its Origin, History and Constitution.* 6th ed. Cambridge University Press, Cambridge.

28. Kahle, C. F (ed.). 1974. *Plate Tectonics: Assessments and Reassessments.* *American Association Petroleum Geologists,* memoir 23. University of Chicago Press, Chicago.

29. Meyerhoff, A. A. and Meyerhoff, H. A. 1974. Tests of plate tectonics. In Kahle, C. F (ed.), *Plate Tectonics: Assessments and Reassessments.* American Association Petroleum Geologists, memoir 23, 43–145. University of Chicago Press, Chicago.

30. Maxwell, J. C. 1974. The new global tectonics: An assessment. In Kahle, C. F (ed.), *Plate Tectonics: Assessments and Reassessments.* American Association Petroleum Geologists, memoir 23, 24–42. University of Chicago Press, Chicago.

31. Vine, F. J. 1966. Spreading of the ocean floor: New evidence. *Science,* v. 154, 1405–1415.

32. Beloussov, V. V. 1974. Seafloor spreading and geologic reality. In Kahle, C. F (ed.), *Plate Tectonics: Assessments and Reassessments.* American Association Petroleum Geologists, memoir 23, 155–166. University of Chicago Press.

33. Beloussov, V. V. 1970. Against the hypothesis of ocean-floor spreading. *Tectonophysics,* v. 9, 489–511.

34. Dewey, J. F. and Bird, J. M. 1970. Mountain belts and the new global tectonics. *Journal of Geophysical Research,* v. 75, 2625–2647.

35. Windley, B. F. 1977. *The Evolving Continents.* Wiley, New York.

36. Basaltic Volcanism Study Project. 1981. Tectonics of Basaltic Volcanism, Chapter 6. In *Basaltic Volcanism on the Terrestrial Planets.* Pergamon, New York.

37. Moores, E. M. and Vine, F. J. 1971. The Troodos Massif, Cyprus and other ophiolites as oceanic crust: Evaluation and implications. *Philosophical Transactions of Royal Society A,* v. 268, 443–467.

38. Condie, K. C. 1997. *Plate Tectonics and Crustal Evolution.* Elsevier, Oxford.

39. Hatherton, T. and Dickinson, W. R. 1969. The relationship between andesitic volcanism and seismicity in Indonesia, the Lesser Antilles, and other island arcs. *Journal of Geophysical Research* v. 74, 5301–5310.

40. Ringwood, A. E. 1974. The petrological evolution of island arc systems. *Journal of Geological Society of London,* v. 130, 183–204.

41. Chappell, B. W. and White, A. J. R. 1974. Two contrasting granite types. *Pacific Geology,* v. 8, 173–174.

42. Pitcher, W. S. 1982. Granite type and tectonic environment. In *Mountain Building Processes* (Hsu, K. J., ed.), 19–40, Academic Press, Oxford.

43. Kuhn, T. S. 1962. *The Structure of Scientific Revolutions.* University of Chicago Press, Chicago.

8   Isotope and Trace Element Geology

INTRODUCTION

In Chapter 2 we outlined the discovery of radioactivity by Henri Becquerel and Marie Curie, and also the discovery of the law of radioactive decay by Ernest Rutherford and Frederick Soddy around the turn of the twentieth century. Rutherford was the first to calculate the age of a mineral using the uranium-helium (U-He) method, but because helium is a gas its mobility led to minimum ages only, the uranium-lead (U-Pb) method proved somewhat more reliable as shown by Bertram Boltwood and Arthur Holmes. These early dating methods are referred to as "chemical ages," where only the concentration ratio of daughter and parent elements are needed; these chemical ages preceded more modern ages based on isotopic ratios now called radiometric ages or isotope ages. In Chapter 2 we also saw that in order to measure the abundance of heavy isotopes, such as lead, an improved mass spectrometer was needed, and this crucial step was made by the physicist Alfred O. Nier who made the first precise measurements of lead isotopes in minerals. Using Nier's data, several workers independently attempted to place constraints on the age of the Earth (Chapter 3). It was not until the 1950s that the first reliable age of the Earth was made using the Pb-Pb isochron technique (see Box 3.1).

In this chapter, some of the different radiometric dating techniques used to solve geological problems are outlined with some early examples. In addition, the use of stable isotopes is addressed, but because this field is very large, discussion here is restricted to oxygen and hydrogen stable isotopes. The two main applications of radiometric methods are dating of individual minerals and dating of whole rocks.

The methods discussed here are limited, due to space constraints, to the U-Pb method on zircon, Rb-Sr and Nd-Sm methods on micas and whole rocks, and the K-Ar method on micas (Table 8.1). The more recent $^{40}Ar/^{39}Ar$ method has largely replaced the K-Ar method, but this technique is left to more specialized publications. The Carbon-14 technique is largely the workhorse of archaeologists, being restricted to events less than 50,000 years old, but it proves useful in dating very young geologic events. Lastly, the Rare Earth Elements (REE) are briefly introduced in anticipation of their importance in lunar petrogensis (Chapter 10), and also due to their importance for terrestrial petrology in general.

MINERAL AGES

In the 1960s mineral age studies in the European Alps showed that Rb-Sr ages on white micas were about 8 Ma older than Rb-Sr biotite ages, with Alpine minerals yielding ages in the range of 60 to 15 Ma.[1] Clearly, discordant ages from the same rock could not reflect the true age of the minerals, and as such they are referred to as "apparent" ages. In addition, comparison of K-Ar and Rb-Sr ages on the same biotites were largely in agreement.[2] The authors of these studies concluded that the minerals had cooled through different temperatures in which the isotopic systems became closed, and that the older white mica Rb-Sr ages recorded a higher cooling temperature compared to biotite ages. Furthermore, the cooling temperatures for the K-Ar and Rb-Sr systems in biotite were similar. The biotite ages in the Alps became abruptly younger from old values (350–150 Ma, reflecting basement ages) to younger Alpine ages (60–15 Ma) as the biotite isograd was crossed corresponding to a metamorphic temperature of about 300°C.[2] It became clear that mineral ages were not crystallization ages, but rather reflected resetting of the isotopic clock during metamorphism.

At about the same time, a study examined the K-Ar system in hornblende (a common amphibole mineral) and the Rb-Sr and the K-Ar systems in biotite as a function of distance from a Tertiary aged pluton

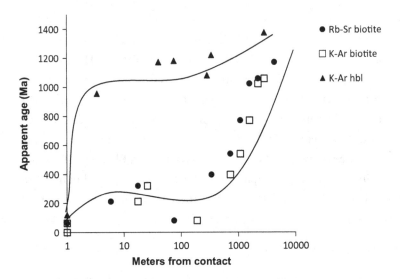

FIGURE 8.1 Apparent mineral ages as a function of distance from the intrusive contact of the Tertiary aged (54 Ma) Eldora pluton into Precambrian country rock, Front Range, Colorado. The Rb-Sr and K-Ar systems on biotite show similar patterns, suggesting similar closure temperatures. The K-Ar pattern in hornblende (hbl) suggests a higher closure temperature for hornblende (Table 8.1).

Source: Data from Hart, S. R. 1964. The petrology and isotopic mineral age relations of a contact zone in the Front Range, *Colorado. Journal of Geology,* v. 72, 493–525. University of Chicago with permission.

(54 Ma) intruded into Precambrian metamorphic rocks (~1400 Ma) in the Front Range of Colorado.[3] This study allowed the effect of temperature due to contact metamorphism to be examined in these mineral systems. The main results are shown in Figure 8.1. The patterns were interpreted in terms of diffusion of the daughter products out of their respective host minerals. The Rb-Sr and K-Ar ages on biotite show somewhat similar patterns where resetting toward younger ages occurs about a thousand meters from the contact, suggesting they have similar cooling temperatures in agreement with observations from the Alps and elsewhere. Hornblende retains its old K-Ar ages up to less than ten meters from the contact, suggesting a higher cooling temperature for this system (Table 8.1).

Table 8.1 *Selected Radiometric Dating Systems*

System	Parent	Daughter	½ life $\times 10^9$ yrs	Mineral	T_c (°C)
U-Pb	^{235}U	^{207}Pb	0.704	Zircon	~900
U-Pb	^{238}U	^{206}Pb	4.47	Zircon	~900
K-Ar	^{40}K	^{40}Ar	1.25	Biotite	~300
				Hornblende	~500
Rb-Sr	^{87}Rb	^{87}Sr	48.8	Biotite	~300
				Muscovite	~500
Sm-Nd	^{147}Sm	^{143}Nd	106	Garnet	~600
C-14	^{14}N	^{12}C	5,700 years	Organics	N/A

The interpretation of mineral ages in terms of cooling ages was put on a quantitative basis with the introduction of the concept of "closure temperature," introduced by the physicist/geologist Martin Dodson in 1973.[4] Dodson proposed that mineral isotopic systems remain open by diffusion of their daughter products at temperatures above the closure temperature, and become closed systems when they pass below that temperature. Because diffusion rates depend exponentially on temperature, as the temperature falls the mineral becomes a closed system over a narrow temperature window. Dodson derived an expression for the closure temperature of a mineral that depends on several variables, including the cooling rate, the diffusion characteristics of the daughter product (diffusion rate and activation energy), the diffusion geometry (dependant on the shape of the mineral), and the grain size of the mineral. This opened up a new field of research, *thermochronology*, whereby the cooling history of mountain belts could be unraveled using mineral ages with different closure temperatures. Some examples of minerals and their closure temperatures (T_c) are given in Table 8.1.

A powerful new U-Pb graphical technique was introduced by George W. Wetherill in the 1950s that had profound implications for the geological study of complex metamorphic terrains common to most ancient mountain belts.[5] As noted previously in Chapter 3, the U-Pb system has two independent decay schemes: namely ^{238}U decays to ^{206}Pb and ^{235}U decays to ^{207}Pb, providing two independent chronometers with very different half-lives (Table 8.1). If the two ages are the same on a mineral within analytical error, they are referred to as concordant ages and they will plot on a concordia curve on a ^{206}Pb/^{238}U vs. ^{207}Pb/^{235}U diagram. Furthermore, Wetherill showed that if the ages are discordant, because of a subsequent resetting event, the original crystallization age of the mineral can still be obtained, and that the age of the resetting event could also be determined. Some ambiguity however still existed; the open-system behavior may be interpreted as due to either continuous diffusion of Pb out of the system (or uranium into the system) or a discrete metamorphic event. An example of this ambiguous situation was presented by E. J. Catanzaro in 1963.[6] Subsequent to Stanley Hart's K-Ar and Rb-Sr mineral work at the Massachusetts Institute of Technology,[3] new work outlined the effect of the Eldora contact metamorphism on resetting of the U-Pb systematics in zircon ($ZrSiO_4$) as a function of distance from the intrusive contact in the Front Range (Figure 8.2).[7] Two samples of country rock zircon (15 meters and 4,267 meters from the contact) both plot on a line (or chord) with an upper intercept of about 1,400 Ma, similar to the age of the country rock and two samples close to the contact (60 centimeters and 3.6 meters) plot close to the lower intercept of the chord with an age similar to that of the Eldora pluton. Examination of the morphology of the zircons showed that the country rock zircons far from the contact were rounded and anhedral in shape, whereas the zircons close to the contact were euhedral suggesting recrystallization or neomineralization of the zircons due to metamorphism. The U-Pb concordia diagram is one of the most useful systems in all of geochronology because of the ability to retrieve both the metamorphic age and the original crystallization age under suitable conditions.

The oldest mineral dated so far is a zircon found in a sandstone from the Jackson Hills in western Australia.[8] Three U-Pb analyses of the zircon's center plot on Wetherill's concordia curve at 4,400 Ma

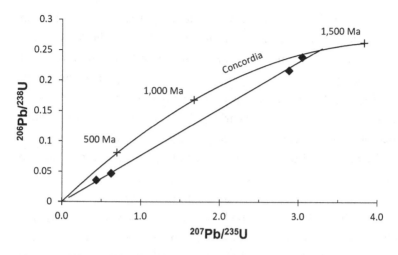

FIGURE 8.2 A U-Pb concordia diagram for zircons from the country rock as a function of distance from the Eldora intrusive pluton (54 Ma), Front Range, Colorado. The chord is defined by four discordant zircons, two close to the contact and two further away. The upper intersect, about 1,400 Ma, corresponds to the age of the country rock, and the lower intercept is similar to the age of the Eldora pluton (54 Ma).
Source: Data from Davis, G. L., Hart, S. R., and Tilton, G. R. 1968. Some effects of contact metamorphism on zircon ages. *Earth Planetary Science Letters*, v. 5, 27–34. Courtesy Elsevier.

(4.4 Ga), which is only 100 Ma younger than the formation of the Earth (i.e., 4.5 Ga). The lower intercept on the concordia curve was at about 3,400 Ma (3.4 Ga), indicating a metamorphic event at that time. It is highly unlikely an older mineral than this will ever be found.

An example of resetting of mineral ages on a mountain-belt scale is represented in the northern Appalachians where K-Ar micas give Permian to Triassic ages over a very large area.[9] The northern Appalachians are polyorogenic with Lower Palaezoic (Taconic), Middle Paleozoic (Acadian), and Upper Paleozoic (Alleghanian) orogenic events overprinting each other. Only in the southeastern part of the region (Rhode Island and Connecticut) can Alleghanian plutonism and high-grade metamorphism be documented because of the presence of Carboniferous rocks. Outside of this relatively small area, whether the reset mineral ages were due to uplift and cooling through a closure temperature or due to the dynamic effects of the Alleghanian metamorphism was unclear. On a U-Pb concordia diagram, many of the zircons from basement rocks in the southern part of the region also display Permian-aged lower intercepts. An important lesson learned from these and other early mineral studies is that the significance of reset mineral ages is commonly equivocal unless detailed morphological and textural studies of the minerals involved are also undertaken at the same time.[10]

Perhaps the best known dating method known to the general public is the carbon-14 method developed by Willard F. Libby and colleagues at the University of Chicago in the 1950s, a method that continues to be mainly used by archaeologists today. This method was first successfully tested on organic objects of known age, such as a 3,000 year old redwood tree and wood from an Egyptian tomb that was about 5,000 years old. Libby received the Nobel prize in chemistry in 1960 for this work.[11]

Most carbon (99%) has an atomic weight of 12 units (^{12}C), but a small amount of carbon is heavier with a weight of 14 (^{14}C) – which is radioactive. This carbon is produced by interaction of cosmic rays with nitrogen in the atmosphere and the carbon-14 produced then begins to slowly decay back to nitrogen over time with a half-life of about 5,700 years. Trees and animals absorb carbon-14 from the atmosphere and the hydrosphere. While alive, the carbon-14 reaches a constant level in living organisms because it is continually being absorbed and decaying at the same time. However, when the plant or animal dies, it stops absorbing the carbon-14 and begins to decay without being replenished. By measuring the amount of carbon-14 in a piece of wood or animal bone or shell, the length of time since the death of the animal or plant can be made. This is the basis of the carbon-14 dating technique, which is only useful to about 40,000 years back. In conventional carbon-14 dating, the amount of carbon-14 is measured by counting its radioactive emissions (beta particles), which requires a large amount of sample (several grams). In the past few decades, a new technique called accelerator mass spectrometry (AMS) has been developed whereby the different carbon isotopes (^{12}C, ^{13}C, and ^{14}C) are measured in a mass spectrometer and the analysis can be performed on very small samples (micrograms).

WHOLE ROCK AGES

The Rb-Sr whole rock system of dating has been the work horse of the geological community since the 1960s. It was recognized in the late 1930s that the decay of rubidium to strontium (^{87}Rb → ^{87}Sr) could be used to date minerals rich in rubidium. But it took the development of the modern mass spectrometer to make dating of common minerals, such as mica and feldspar, possible by this method.[12] Researchers at Oxford University identified some of the oldest rocks in the world, at that time, from coastal Greenland using the Rb-Sr isochron method (Figure 8.3).[13] This method assumes that a suite of co-magmatic rocks share the same initial strontium isotopic composition (^{87}Sr/^{86}Sr)$_i$. The decay of ^{87}Rb to ^{87}Sr increases the ^{87}Sr/^{86}Sr ratio with time, and this ratio increases in proportion to the Rb/Sr ratio of the rock so that a co-magmatic suite of rocks fall on a straight line on a plot of ^{87}Sr/^{86}Sr versus ^{87}Rb/^{86}Sr. The line (y = mx +c) can be expressed as:

$$^{87}Sr/^{86}Sr = (^{87}Rb/^{86}Sr)\lambda t + (^{87}Sr)_i$$

where λ is the half life and t is time (Table 8.1). This graphical method (Figure 8.3) of displaying Rb-Sr whole rock results was first suggested by L. O. Nicolaysen in 1961.[14] The isochron line is fitted to the whole rock data points using a least-squares statistical method, and the slope determines the age of the suite of rocks:

$$Age = 1/\lambda \ln (slope + 1).$$

It is quite common for individual minerals in a rock to be reset during metamorphism, as described earlier for the examples in Colorado and the Alps. The most valuable characteristic of the whole rock system is that the whole rock isochron commonly preserves the original crystallization age.

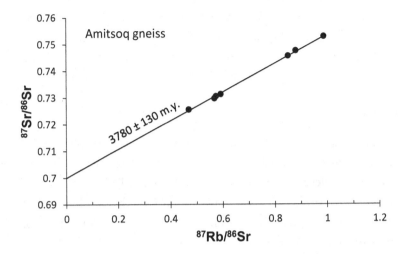

FIGURE 8.3 A Rb-Sr whole rock isochron for the Amitsoq gneisses, Isua, western Greenland. These tonalitic gneisses have an early Archean age (3,780 Ma) and were amongst the oldest rocks known at the time. The low initial $^{87}Sr/^{86}Sr$ ratio of 0.699 (similar to that of meteorites) indicates a primitive mantle origin.

Source: Data from Moorbath, S., O'Nions, R. K., and Pankhurst, R. J. 1975. The evolution of early Precambrian crustal rocks at Isua, west Greenland – geochemical and isotopic evidence. *Earth Planetary Science Letters*, v. 27, 229–239. Courtesy Elsevier.

In addition, the intersection of the isochron line with the ordinate axis on an isochron plot gives the strontium isotopic composition of the rock suite at the time of its formation (the initial $^{87}Sr/^{86}Sr$ ratio), which itself is a powerful chemical tracer in terms of magmatic origin. In Chapter 7 we mentioned a classification of granites into I-type (igneous) and S-type (sedimentary). In addition to their mineralogical and chemical characteristics, these two granite types can also be distinguished by their initial $^{87}Sr/^{86}Sr$ ratios: >0.707 for the S-type, and <0.707 for the I-type.[15] In the Sierra Nevada granite batholith of the western United States, this ratio varies from 0.704 in the west to 0.708 in the east, likely reflecting a mantle source in the east and a crustal source farther inland.[16]

That the continental crust was extracted from the Earth's upper mantle over geologic time can be seen by examining the initial

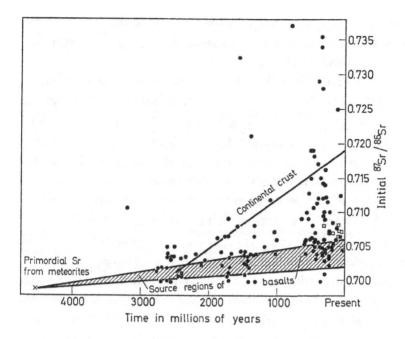

FIGURE 8.4 A plot of the initial $^{87}Sr/^{86}Sr$ ratio of granitic rock against time. The radiogenic strontium initial ratios are based on Rb-Sr whole rock isochrons for each sample suite (dots). The shaded region represents the upper mantle based on the initial ratio of oceanic basalts. The initial radiogenic strontium is that of meteorites at 4,500 Ma. Approximately half of the granites plotted in this region indicating a mantle source. The continental crust line assumes a mantle source with a Rb/Sr ratio of 0.18 and an age of 2,500 Ma. This is just one example of many such lines that could be drawn. Granites with high initial ratios (>0.71) are probably derived from remelting of the continental crust. Open boxes are for North American batholiths. The data compilation is based on the literature up to 1971.

Source: Faure, G. and Powell, J. L. 1972. *Strontium Isotope Geology*. Springer-Verlag, New York. Courtesy Springer.

strontium isotopic ratio of granites as a function of time (Figure 8.4).[12] This extraction left behind depleted upper mantle that is the source of mid-ocean ridge basalts (MORB). In the case of modern volcanic rocks (whose geologic age is zero), the measured $^{87}Sr/^{86}Sr$ ratio is also the initial value; the same concept can also be applied to these rocks, allowing ocean floor, ocean island, and island arc basaltic rocks to be distinguished by their measured $^{87}Sr/^{86}Sr$ ratio, although their ranges overlap.[17]

The samarium and neodymium whole rock system ($^{147}Sm \rightarrow$ ^{143}Nd) is a more recent development compared to the Rb-Sr system, largely because of the very long half-life of ^{147}Sm (Table 8.1). A long half-life indicates that small radiogenic changes are produced requiring high precision mass spectrometry in addition to very low laboratory contamination levels. The first Sm-Nd ages were determined on meteorites by Gunter W. Lugmair at the University of California, in 1974.[18] The Sm-Nd parent-daughter pair belong to rare earth group of elements (also referred to as the lanthanide elements by chemists), and this decay system is graphically identical to the Rb-Sr whole rock isochron system with the $^{147}Sm/^{144}Nd$ ratio representing the abscissa and the $^{143}Nd/^{144}Nd$ ratio plotted on the ordinate. In this case ^{144}Nd is the stable isotope, analogous to ^{86}Sr in case of the Rb-Sr system. Samarium and neodymium, however, have distinctly different chemical behaviors such that the Sm/Nd ratio *decreases* from mafic to felsic rocks, whereas the Rb/Sr ratio increases in this direction. The Sm-Nd system is especially useful in dating mafic rocks, where the Sm/Nd ratio and hence radiogenic Nd tends to be higher. An example of a Sm-Nd isochron for an Apollo 15 breccia fragment of gabbro, which apparently dates the lunar crust, is given in Chapter 10. Because the rare earth elements also tend to be immobile, the Sm-Nd method can withstand the effects of metamorphism better than the Rb-Sr system.

The radiogenic strontium and radiogenic neodymium ratios can be used in tandem to examine the origins of I-type and S-type granites. As mentioned, S-type granites tend to have higher radiogenic strontium ratios, and correspondingly they tend to have lower radiogenic neodymium ratios; the reverse situation is true for I-type granites. These ratios are plotted for I-type and S-type granites from southeastern Australia, where they appear on a mixing line between a mantle component and a crustal component (Figure. 8.5).[19]

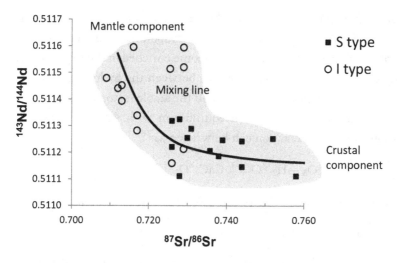

FIGURE 8.5 Plot of $^{143}Nd/^{144}Nd$ versus $^{87}Sr/^{86}Sr$ for I-type and S-type granites from southeastern Australia. The samples lie along a mixing line between a mantle component (high radiogenic neodymium and low radiogenic strontium) and a crustal component (high radiogenic strontium and low radiogenic neodymium) sources. This is consistent with mineralogical and chemical characteristics of these granite types.
Source: Data from McCulloch, M. T. and Chappell, B. W. 1982. Nd isotopic characteristics of S and I-type granites. *Earth Planetary Science Letters*, v. 58, 51–64. Courtesy Elsevier.

STABLE ISOTOPES

Oxygen and hydrogen both occur in the atmosphere, hydrosphere, biosphere, and the lithosphere.[12] This section introduces oxygen and hydrogen isotopes and their application to paleoclimate studies, which is treated in the next chapter in the context of the Pleistocene Ice Age. Oxygen is the most abundant element on Earth and its two most abundant isotopes are ^{16}O (99.6%) and ^{18}O (0.25%). The two isotopes of hydrogen are ^{1}H and ^{2}H, or deuterium (D), first identified by Harold C. Urey at the University of Chicago in 1932. Ordinary water is H_2O, and so-called heavy water, used in some types of nuclear reactors, is D_2O. There is a tendency for the lighter isotopes to form weaker bonds so that during processes, such as evaporation, the isotopes tend to fractionate from each other, and then fractionation also increases with decreasing temperature. In the case of evaporation, for example, the lighter isotopes ^{16}O and H tend to become enriched in the vapor phase. Because of the large mass difference between the hydrogen isotopes, they show the greatest variation of all the stable isotopes.

In the case of calcite in equilibrium with water, the oxygen isotopic exchange reaction can be written as:[16]

$$\tfrac{1}{3}CaC^{16}O_3 + H_2{}^{18}O = \tfrac{1}{3}CaC^{18}O_3 + H_2{}^{16}O.$$

The fractionation factor between calcite and water is defined as:

$$\alpha = R_c/R_w,$$

where $R_c = {}^{18}O/{}^{16}O$ in calcite and $R_w = {}^{18}O/{}^{16}O$ in water. The delta (δ) notation expresses the per mil (one in a thousand) difference between the isotopic composition of a sample and a standard:

$$\delta^{18}O = \left[\left({}^{18}O/{}^{16}O_{samp} - {}^{18}O/{}^{16}O_{std}\right)/\left({}^{18}O/{}^{16}O_{std}\right)\right] \times 1,000,$$

and an analogous expression can be written for delta deuterium (δD). A commonly used standard is standard mean ocean water (SMOW). Positive values of $\delta^{18}O$ and δD indicate the sample is enriched in ^{18}O and D relative to the standard, and negative values indicate it is depleted in these isotopes. Because the fractionations of hydrogen and oxygen isotopes are proportional to each other as a function of temperature, a linear relationship exists between δD and $\delta^{18}O$ values in the case of meteoric water as a function of latitude (Figure 8.6).[20]

As already noted, during evaporation the vapor phase is enriched in the lighter isotopes (^{16}O and H). If precipitation from a moist air mass occurs either as rain or snow, the heavier isotopes will be preferentially precipitated leaving the cloud phase further enriched in the lighter isotopes. As the clouds move to higher latitudes over the continents, the temperature decreases and the fractionation of the isotopes also increases such that precipitation at high latitudes has very negative δD and $\delta^{18}O$ values. Ice sheets accumulated at these

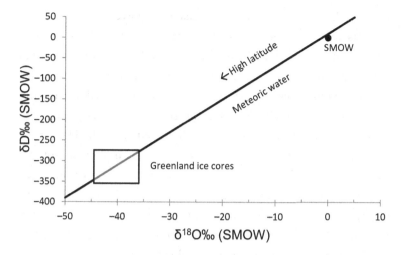

FIGURE 8.6 Plot of δD versus $\delta^{18}O$ for meteoric water. These ratios become more negative at high latitudes (e.g., Greenland and Antarctica). The composition of ice from Greenland ice cores is indicated. These cores record climate changes going back about 100,000 years. SMOW: standard mean ocean water. The equation of the straight line is δD = 8 $\delta^{18}O$ + 10.

Sources: Craig, H. 1961. Isotopic variations in meteoric waters. Science, v. 133, 1702–1703, and NOAA. American Association for Advancement of Science with permission.

localities (e.g., Greenland or Antarctica) over time will record these low isotopic values reflecting the local ground temperature.

Because of the temperature dependence of isotope fractionation between phases, it was suggested by Harold Urey at the University of Chicago in the 1940s that the $\delta^{18}O$ values of $CaCO_3$ from microfossil marine shells (forams) could be used to estimate the temperature of coexisting sea water. Subsequent experimental work provided an empirical relationship between temperature and the $\delta^{18}O$ values of water in equilibrium with carbonate.[21] Before this project could be undertaken, however, several major technical obstacles had to be overcome. The small size of the foram samples (fractions of a milligram) meant a more precise mass spectrometer had to be designed to analyze such small samples and to detect very small isotopic variations. Chemical procedures also had to be developed that would yield consistent results. These problems were solved by various research groups under Urey's direction at the University of Chicago in the 1950s. In practice, the calcium carbonate of the foram shells was converted to carbon dioxide (CO_2) under fixed conditions, and the oxygen isotopic composition of the resulting gas was analyzed.

Shortly thereafter, in 1955, an Italian student at the University of Chicago, Cesare Emiliani, published a large number of isotopic measurements on the calcite shells of foraminifera from deep-sea cores.[22] He interpreted the large cyclic variations in $\delta^{18}O$ as due to variations in the ocean temperature (estimated to have been as much as 6°C) during Pleistocene glacial and interglacial cycles. Later it was concluded that the variations were less due to temperature variations than to changes in the isotopic composition of the ocean water itself. Emiliani's higher $\delta^{18}O$ values from deep-sea cores therefore actually reflected the volume of ice on the continents during glacial periods, but it would be some time before this conclusion was reached. These results are discussed in more detail in the next chapter.

RARE EARTH ELEMENTS

It was mentioned above in the context of the Sm-Nd sytem of radiometric dating that these elements belong to the rare earth group of elements (REE). This group of elements is the most useful of the trace elements in interpreting igneous, metamorphic, and sedimentary processes.[23,24] They comprise the elements from atomic number 57 to 71 – lanthanum to lutetium (Table 8.2). Their most important characteristic is that the radius of these ions (most in the +3 state, except for europium which can also be Eu^{2+}) progressively decreases with increasing atomic number. The REE patterns, therefore, show smooth patterns that can be used to infer petrogenetic processes. Note in Table 8.2 the big difference in the ionic radius of europium in the +3 state and the +2 state. This can lead to positive and negative europium anomalies in REE patterns if plagioclase feldspar is involved in the evolution of the igneous rocks in question (Chapter 10). In addition,

Table 8.2 *Rare Earth Elements*

Atomic number	Element	Symbol	Ionic radius (Å)* (+3)
57	Lanthanum	La	1.160
58	Cerium	Ce	1.143
59	Praesodymium	Pr	1.126
60	Neodymium	Nd	1.109
61	Promethium	Pr	Short-lived
62	Samarium	Sm	1.079
63	Europium	Eu	1.066 (+3)
			1.250 (+2)
64	Gadolinium	Gd	1.053
65	Terbium	Tb	1.040
66	Dysprosium	Dy	1.027
67	Holmium	Ho	1.015
68	Erium	Er	1.004
69	Thulium	Tm	0.994
70	Ytterbium	Yb	0.985
71	Lutetium	Lu	0.977

* Angstroms (10^{-10} m). Source of ionic radii: Shannon, R. D. 1976. Revised effective ionic radii and systematic studies of interatomic distances in halides and chalcogenides. *Acta. Crystallographia, ser. A*, v. 32, 751–767. Courtesy of the International Union of crystallography.

quantitative modeling of the REE concentrations can be used to test different petrogenetic hypotheses.[25] The lower atomic members of the series are referred to as light rare earths (LREE) and the higher atomic numbers are heavy rare earths (HREE). Their concentrations are usually normalized to stony meteorite concentrations and plotted on a log scale.

The field of using the REE to solve petrogenetic problems began in the 1970s, and the Mineralogical Society of America reviewed the field in a 1989 volume.[25]

The formation of the continents has to be one of the most important events in geologic history. Figure 8.4 is a plot of the initial $^{87}Sr/^{86}Sr$ ratio for granites plotted against time with the field of the

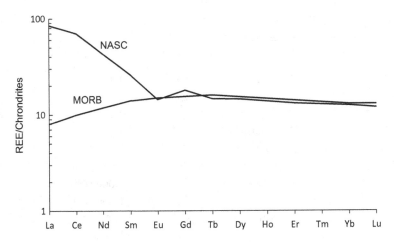

FIGURE 8.7 A plot of rare earth element (REE) patterns for a composite of
North American shales (NASC) and an average of mid-ocean ridge basalts
(MORB) normalized to chrondritic meteorites. The NASC pattern is
thought to reflect the average upper continental crust and is strongly
LREE enriched. The MORB pattern is LREE depleted, reflecting their
depleted upper mantle source. The two patterns are complementary in
the LREE region and reflect the extraction of the continental crust from
the upper mantle over time.
Sources: Data from Gromet, L. P. and others. 1984. The North American shale
composite: its compilation, major and trace element characteristics. *Geochemica
Cosmochimica Acta*, v. 48, 2469–2482; Basaltic Volcanism Study Project. 1981.
Basaltic Volcanism on the Terrestrial Planets. Pergamon. New York. Courtesy
Elsevier and The Lunar and Planetary Institute.

upper mantle also shown. The inference from this plot is that approximately 50% of granites were extracted from the upper mantle during
geologic time, and the remainder are the result of melting within the
continental crust. Comparison of the REE patterns of the average
upper continental crust and that of average mid-ocean ridge basalts
(MORB) yields a similar conclusion. Because the LREE are relatively
large, they are incompatible with respect to major rock-forming minerals so that they are partitioned preferentially into the late (granitic)
fraction during melting. This results in depletion of the upper mantle
in the LREE (and enrichment in the continents) over geologic time.
The production of mid-ocean ridge basalt by partial melting of the

upper mantle therefore should show a LREE depleted pattern reflecting its source. The REE element pattern for a composite sample of North American Shales (NASC)[26] is thought to be representative of the average upper continental crust and it is plotted along with the average mid-ocean ridge basalt in Figure 8.7.[27] The two patterns are identical in the HREE region at about ten times stony meteorites, but are complementary to each other in the case of the LREE, consistent with expectation. In Chapter 10 we will see a somewhat analogous situation for the REE patterns of lunar mare basalts and of the lunar highland crust.

REFERENCES

1. Jäger, E. 1962. Rb-Sr determinations on micas and total rocks from the Alps. *Journal of Geophysical Research*, v. 67, 5293–5306.
2. Armstrong, R., Jäger, E., and Eberhardt, P. 1966. A comparison of K-Ar and Rb-Sr ages on Alpine biotites. *Earth Planetary Science Letters*, v. 1, 13–19.
3. Hart, S. R. 1964. The petrology and isotopic mineral age relations of a contact zone in the Front Range, Colorado. *Journal of Geology*, v. 72, 493–525.
4. Dodson, M. 1973. Closure temperature in cooling geochronological systems. *Contributions in Mineralogy and Petrology*, v. 40, 259–274.
5. Wetherill, G. W. 1956. Discordant uranium-lead ages. *Transactions of American Geophysics Union*, v. 37, 320–326.
6. Cattanzaro, E. J. 1963. Zircon ages in southwestern Minnesota. *Journal of Geophysical Research*, v. 68, 2045–2048.
7. Davis, G. L., Hart, S. R., and Tilton, G. R. 1968. Some effects of contact metamorphism on zircon ages. *Earth Planetary Science Letters*, v. 5, 27–34.
8. Valley, J. W., Cavosie, A. J., Ushikubo, T. et al. 2014. Hadean age for a post-magma-ocean zircon confirmed by atom-probe tomography. *Nature Geoscience*, v. 7, 219–223.
9. Zartman, R. E., Hurley, P. M., Krueger, H. W. and Giletti, B. J. 1970. A Permian disturbance of K-Ar radiometric ages in New England. *Geological Society America Bulletin*, v. 81, 3359–3374.
10. O'Hara, K. D. and Gromet, L. P. 1983. Textural and Rb-Sr isotopic evidence for late Paleozoic mylonitization within the Honey Hill fault southeastern Connecticut. *American. Journal of Science*, v. 283, 762–779.
11. Libby, W. F. 1980. Archaeology and radiocarbon dating. *Radiocarbon*, v. 22, 1017–1020.

12. Faure, G. and Powell, J. L. 1972. *Strontium Isotope Geology*. Springer-Verlag, New York.

13. Moorbath, S., O'Nions, R. K., and Pankhurst, R. J. 1975. The evolution of early Precambrian crustal rocks at Isua, west Greenland – geochemical and isotopic evidence. *Earth Planetary Science Letters*, v. 27, 229–239.

14. Nicolaysen, L. O. 1961. Graphic interpretation of discordant age measurements on metamorphic rocks. *Annual New York Academy of Sciences*, v. 91, article 2, 198–206.

15. Chappell, B. W. and White, A. J. R. 1974. Two contrasting granite types. *Pacific Geology*, v. 8, 173–174.

16. Kistler, R. W. and Peterman, Z. E., 1973. Variations in Sr, Rb, K, Na and initial $^{87}Sr/^{86}Sr$ in Mesozoic granitic rocks and intruded wall rocks in Central California. *Geological Society America Bulletin*, v. 84, 3489–3512.

17. Faure, G. 1986. *Principles of Isotope Geology*. 2nd ed. Wiley, New York.

18. Lugmair, G. W. 1974. Sm-Nd ages: A new dating method. *Meteoritics*, v. 9, 369.

19. McCulloch, M. T. and Chappell, B. W. 1982. Nd isotopic characteristics of S and I-type granites. *Earth Planetary Science Letters*, v. 58, 51–64.

20. Craig, H. 1961. Isotopic variations in meteoric waters. *Science*, v. 133, 1702–1703.

21. Epstein, S., Buchsbaum, R., Lowenstam, H. A. and Urey, H. C. 1953. Revised carbonate-water isotopic temperature scale. *Geological Society of America Bulletin*, v. 64, 1315–1326.

22. Emiliani, C. 1955. Pleistocene Temperatures. *Journal of Geology*, v. 63, 538–578.

23. Lippin, B. R. and McKay, G. A. (eds.). 1989. *Geochemistry and Mineralogy of the Rare Earth Elements*. Reviews in Mineralogy, v. 21. Mineralogical Society of America, Washington DC.

24. Taylor, S. R. and McLennan, S. M. 1985. *The Continental Crust: Its Composition and Evolution*. Blackwell, London.

25. Hanson, G. H. 1989. An approach to trace element modeling using a simple igneous system as an example. In Lippin, B. R. and McKay, G. A. (eds.), *Geochemistry and Mineralogy of the Rare Earth Elements*. Reviews in Mineralogy v. 21. Mineralogical Society of America, Washington DC.

26. Gromet, L. P., Dymek, R. F., Haskin, L. A. and Korotev, R. L. 1984. The North American shale composite: its compilation, major and trace element characteristics. *Geochemica Cosmochimica Acta*, v. 48, 2469–2482.

27. Basaltic Volcanism Study Project. 1981. *Basaltic Volcanism on the Terrestrial Planets*. Pergamon, New York.

9 Ice Ages and Ice Cores

> *We now know, from actual observation, the limits and prominent characteristics of the glaciated area on this continent.*
>
> – G. F. Wright, 1889[15]

GLACIAL DEPOSITS

The Giétro glacier on the north side of the Pennine Alps in the canton of Valais, Switzerland, already had a history of causing destruction in the past. In 1818, in the Bagnes Valley, the growing glacier blocked a stream until it became a two-kilometer-long lake (Figure 9.1). The local civil engineer, Ignatz Venetz (1788–1859), was asked by the canton to help prevent another flooding of the village downstream.[1] Venetz attempted to cut a channel in the ice to drain the lake gradually, but the ice failed catastrophically, causing forty-four deaths and great loss of property.

Venetz continued to study glaciers in the Bagnes Valley and he noted that striations in the rocks produced by glaciers extended well down below the valley, concluding that the glaciers were previously more extensive. He also attributed glacial erratics, large exotic boulders whose source was not local, to the work of ice, rather than flowing water as was the opinion at that time. He presented his findings in 1821 to the Swiss Natural Science Society, which had offered a prize on the topic of climate deterioration, that he won.[1]

His ideas were largely ignored until a friend, Jean de Charpentier, sent his work to Paris for publication in 1833, where it gained wide recognition in Europe. Charpentier (1786–1855) was a mining geologist and a student of Abraham Werner at the mining school in Freiberg, Germany. He became director of salt mines in the canton of Vaud, Switzerland, and after the disaster of 1818 he began to study

FIGURE 9.1 Painting of the Giétro glacier, Bagnes Valley, Switzerland, by the Alpine geologist Arnold Escher von der Linth (1807–1872). The painting is dated to 1818, the year the glacier failed and caused serious loss of life and property downstream.
Source: Wikipedia.

glaciers and determined that Venetz was probably correct in that the Alpine glaciers were more extensive in the past.[2] He studied the Rhône Valley and the lowland region between the Alps and the Jura mountains for evidence of glaciers, and he concluded that the entire region had been covered by a mega-glacier (*glacier-monstre*)[3]. In 1835 he published his paper on the transport of erratics in the Rhône Valley and his more extensive book *Essai sur les glaciers* (Essay on glaciers) was published in 1841.[4]

In October 1838, the clergyman William Buckland and his wife crossed the English Channel by steamer and then traveled by horse and carriage to visit Louis Agassiz, then a young professor at Neuchâtel, Switzerland (we met both Buckland and Agassiz in Chapter 1). Agassiz had presented his ideas on glaciers, moraines, and erratic boulders to the Swiss Society of Natural Sciences a year earlier.[5] Buckland and Agassiz were both already friends as Agassiz had previously studied fossil fish in England with the help of Buckland, who had procured funding for him from the British Science Association.[6] Having changed his research focus from fossil fish (Chapter 1) to glacial processes since arriving at Neuchâtel, Agassiz's goal was to persuade some of the most respected geologists in northern Europe that there had been an Ice Age represented by sediments previously thought by Buckland and many others to represent the biblical flood (so called diluvial sediments). Lyell referred to this as "diluvial humbug" in a 1829 letter to a friend. Glacial features such as striated bedrock and glacial erratics are difficult to explain by the action of water, but that was the most common explanation at the time. Charles Lyell had an alternative explanation: he attributed these erratics to drifting icebergs that deposited their cargo on melting; this theory allowed him to maintain his uniformity principle so that catastrophic floods were not required.[7] He suggested that sediments deposited in this manner be called "drift" and glacial sediments are still referred to as drift on British Survey geologic maps. Modern studies of Atlantic deep sea cores show that drifting icebergs did indeed deposit thin layers of sand on the ocean floor, but this was never a plausible explanation for the extensive layers of thick drift deposits on the continents in the northern hemisphere.

A second distinctive type of glacially derived sediment is called *loess* (German for loose; it is pronounced like it rhymes with bus), which is a yellowish silt that was produced by the grinding action of the ice over bedrock. This silt was deposited in the outwash plains in front of melting glaciers and later became redeposited by wind to cover large areas of Europe, Asia, and North America.[8] The Yellow River of northern China gets its name for the large quantities of easily eroded loess that the river carries downstream.

It also makes fertile soil for the bread basket regions of the Ukraine and the midwest of the United States.

Louis Agassiz led a field trip in the Jura mountains near Neuchâtel with Buckland as his guest in 1838 and showed him features in the rocks and sediments that indicated a previous glacial episode including striated bedrock, glacial erratic boulders (*bloc erratiques*), and glacial moraines. They also examined the glaciers themselves at higher elevations, where Buckland remarked that he had seen similar features in Scotland and so was persuaded by Agassiz's theory.[6] At the annual presidential address of the Geological Society of London in 1841, Agassiz presented his evidence for a glacial episode in Scotland, England, and Ireland based on his own field work.[9] He first gave credit in his address to Venetz and Charpentier, who recognized the widespread action of ice in the Alps and surrounding regions. This address was followed by Buckland himself with evidence in areas of Scotland and England where he had observed similar glacial features described by Agassiz.[10] Buckland had already persuaded Lyell of the importance of these features in Scotland, as they indicated a glacial period in the past.

Agassiz had now succeeded in his goal of persuading the scientific elite, who had rejected his theory up to that time, that sheets of thick ice (*nappes* as Agassiz called them, not to be confused with tectonic nappes) had capped northern Europe in the recent past. His major work *Etudes sur les glaciers* (Studies of glaciers) published in 1840 goes further than either Venetz or Charpentier – they interpreted evidence for glacial activity in the Alps to be just an extension of the present day glaciers. Agassiz had a far bolder notion, namely that all of northern Europe at a latitude north of 35° N was covered by a thick sheet of ice at a time earlier than that indicated by the present glaciers. This earlier epoch would later be called the Pleistocene by Lyell. Agassiz was clearly a young professor in a hurry – his book of 1840 was published before Charpentier's book of 1841, even though it was Charpentier who introduced Agassiz to glacial geology.[6]

The Scottish amateur geologist Charles Maclaren and editor of the *Scotsman* summarized Agassiz's book in 1842 to a wider American audience.[11] In that summary, Maclaren appears to be the first to infer that if Agassiz's theory was correct that a large drop in sea level would have occurred; Maclaren estimated that drop to be between 350 and 400 feet (~100–120 meters). In 1868 an American geologist from Ohio, Charles Whittlesey, made a similar estimate based on a more extensive geographic analysis.[12] Modern estimates based on dating of coral reef terraces and isotope stratigraphy indicate the sea level was ~120 meters below current sea level about 17,000 years ago, at the end of the Last Glacial Maximum, confirming the early estimates.[13] This sea level drop allowed a land bridge to form across the Bering Strait, allowing humans to occupy Alaska from Eurasia at this time – possibly for the first time.[14]

It took only twenty years for the geological community in Europe as a whole to accept the concept of a recent Ice Age; Murchinson (see Chapter 1) was a notable exception who rejected the glacial theory until the end. As noted previously, after the death of Agassiz's wife in 1847, he accepted a professorship at Harvard where he continued to teach until 1873. During his career at Harvard, he traveled widely and showed that North America also underwent an Ice Age similar to that in Europe (Figure. 9.2). It is now common knowledge that vast areas of the landscape of northern Europe and North America bear the unmistakable markings of a past Ice Age, including: U-shaped valleys, cirques, polished and striated bedrock, glacial erratics, drumlins, and moraines, to name a few. The Great Lakes of Canada and the United States are also a product of the last glaciation.[15] The precursor to the modern Great Salt Lake was a lake that covered most of Utah, named Lake Bonneville, which drained north into the Columbia River during the Pleistocene.[16]

Much of the subsequent research in glacial geology focused on unraveling the several cycles of glaciation seen in the continental sedimentary record as provided by terminal and recessional moraines,

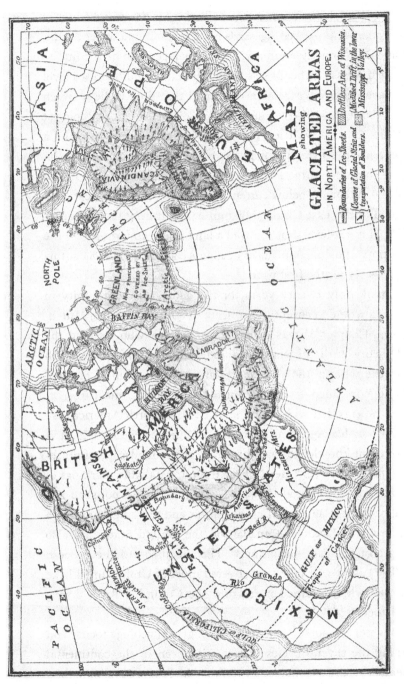

FIGURE 9.2 Extent of continental ice in North America and Europe during the last glaciation, as indicated by G. Frederick. Wright in 1890. Arrows indicate glacial striations and transportation of erratic boulders.

Source: Wright, G. F. 1890. *The Ice Age in North America: And its bearings upon the antiquity of man.* Appleton & Co., New York.

indicating recession and advance of the ice. James Geikie, brother of Archibald, recognized that in Scotland the drift (glacial deposits) was commonly separated by plant-bearing soils, implying that there were several glacial epochs separated by warmer interglacials. Several major glacial advances were identified in Europe (from youngest to oldest: Würm, Riss, Mindel, Gunz, and Donau) and in North America these were called Wisconsin, Illinoian, Kansan, and Nebraskan. Because advancing glaciers tended to destroy preexisting moraines, the continental sedimentary record of the Pleistocene glaciation is very incomplete. In the 1970s deep-sea sediment cores provided much more complete records compared to the continental record, and, more recently still, ice cores from polar regions have corroborated the deep-sea record with still more detailed records.

The idea that ice ages were caused by variations in the Earth's orbit around the sun was proposed by John Herschel in 1830 and further developed by James Croll (1821–1890), a self-taught Scotsman who could not afford university. He spent much of his early career in manual labor occupations, but when his health failed he turned to business as a tea shop owner and insurance salesman – but he was unsuccessful at both.[8] He eventually found a low-level position in a museum with access to a library. He found time to read widely in the sciences and became interested in the cause of ice ages, a topic of general interest since Agassiz's theory became widely known. He published an academic paper on orbital variations of the Earth as a cause of ice ages in 1864.[17] In that paper, Croll proposed that the eccentricity of the Earth's orbit about the sun changes with time, resulting in more or less solar energy reaching the Earth. But he saw that eccentricity alone was insufficient to produce a significant change in climate; he further proposed that the precession of the equinoxes together with eccentricity could cause an ice age. In addition he correctly recognized the importance of ocean currents, such as the Gulf Stream, in carrying heat from the equator to the poles and that these currents would amplify orbital effects during an ice age. Furthermore, he recognized an important positive feedback mechanism in climate change — that snow and ice cover would reflect solar radiation back into the atmosphere, causing further cooling and growth of the ice and so magnifying the initial effect. The reverse effect is seen today as Arctic ice melts, causing the ocean to warm causing further melting and so on.

His 1864 paper drew the attention of Archibald Geikie, then the director of the Geological Survey, who offered Croll a position in the Edinburgh office of the survey in 1867, where he stayed until 1880.[8] Croll is best known for his book *Climate and Time*, published in 1875, which summarized his previous publications and greatly influenced subsequent scientists who tackled the Ice Age problem.[18] During his lifetime, Croll's theory was very controversial – he predicted that the last glacial episode occurred about 100,000 years ago, but at the time geologists estimated the age of the drift deposits as being only about 10,000 years old.

A Yugoslavian engineer and mathematician Milutin Milankovitch (1879–1958) joined the faculty at the University of Belgrade in 1909 and turned his interests toward the cause of ice ages. He set out to calculate the amount of solar radiation that reached the Earth in the past at different latitudes.[19] With the help of the Russian climatologist Wladimir Köppen (who, incidentally, was related to Alfred Wegener through marriage), he inferred that the amount of sunshine on the continents at high latitudes in the summer would control the advance and retreat of ice sheets[20]. The reasoning was as follows: If it becomes colder during the winter, things don't change much because it is already cold. However, if it becomes colder during the summer, less ice will melt and during the following winter, the glaciers will grow larger. Milankovitch then set about calculating the changes in heat from the sun due to orbital variations of the Earth around the sun at 45°N latitude in the summertime.[20] All his calculations were done with pen and paper. He later recalled that at one point he worked all day every day for one hundred days without a break.[8] Some of his calculations were done while he was a prisoner of war in the Balkans for a short time during World War I.

Milankovitch already knew, partly from Croll's work, that three orbital variations affected the amount of solar heat that reached the Earth over time:

- Variations in the shape of the Earth's orbit around the sun – termed its *eccentricity*. This variation has a periodicity of 100,000 years.
- Variations in the tilt of the Earth's axis. This variation (21.5°–24.5°) has a periodicity of 41,000 years. This tilt is the cause of our seasons, and is currently 23.5°.
- The wobble (or precession of the equinoxes) of the Earth's axis, which has a period of 23,000 years.

At certain times the three periodic changes coincide, causing peaks and troughs in solar radiation to reach the Earth. The results of his calculations, which were a series of sine-like curves for solar heating over the past 600,000 years, were broadly consistent with what geologists had known about past ice ages, but there was no way to prove this. The sedimentary record of past ice ages on the continents was too fragmentary, and there was, as yet, no way to date these glacial deposits.

DEEP SEA CORES

The beginning of modern oceanography dates to the research expedition of the British Navy ship *Challenger* (1872–1876), which explored the ocean depths worldwide and collected samples by dredging and trawling for animal and plant life. The results of the research were published in a fifty-volume series between 1880 and 1895 under the supervision of John Murray, one of the scientists aboard the expedition.[21] A common type of sediment observed in the oceans worldwide by the expedition was a calcareous ooze composed of single-celled microscopic animals called foraminifera (forams for short), which include bottom-dwelling (benthic) forms and also shallow-dwelling (planktonic) forms. Different species have different preferences for temperature and salinity, most favoring warm water.

The German research vessel *Meteor*, on an expedition in 1925–1927, raised short (one meter in length) sediment cores from the Atlantic Ocean floor, and the German paleontologist Wolfgang Schott decided to examine the cores for the assemblages of forams present with depth in the cores. He found three different assemblages: the uppermost assemblage was the same as the modern seafloor assemblage and a third deeper assemblage was also similar to today's, but a second intermediate layer was different; a particular warm-water species was absent in this intermediate layer. Schott concluded that the uppermost layer and the third layer represented an interglacial warm epoch, and that the intermediate layer represented the last glacial period. This led to the possibility that longer sediment cores could record climate changes during the entire Pleistocene, a goal that was not reached until several decades later.[8]

The problem of retrieving longer cores was solved by a Swedish oceanographer Björe Kullenberg in 1947. He invented the piston core, in which sediments are sucked into the core as it is driven into the sediment under gravity. The timing of the invention was fortunate, as the Swedish expedition *Albatross* (1947–1948) was the first to use the piston core, and ultimately obtained cores of ten to fifteen meters long from all the major ocean basins. Under the direction of Maurice Ewing, the Lamont Geological Observatory of Columbia University had its own collection of hundreds of cores from around the world. A research group at Lamont studied the cold and warm assemblages of forams in eight widely separated cores from the Atlantic and identified the same abrupt climate change that Schott had observed in his short cores.[22] Crucially, their study also presented carbon-14 dates from the cores, which showed that the abrupt climate change occurred about 11,000 years ago in all eight cores, corresponding to the end of the Pleistocene and the beginning of the present Holocene warm period. The end of the Pleistocene ice age was slowly coming into focus, but the beginning of the Pleistocene had yet to be dated in a deep-sea core.

At about the same time, an Italian student, Cesare Emiliani, came to the University of Chicago to take up the suggestion of Harold Urey, a Nobel laureate in chemistry, that oxygen isotopes could be used to study temperature variations in forams from deep sea cores (Chapter 8). These microscopic plankton sink to the bottom of the ocean floor when they die, but during their lifetime, they built their calcareous shells from calcium and oxygen from surface seawater. The partitioning of oxygen isotopes between seawater and calcite shells is temperature dependant as theoretical and experimental studies had shown, allowing the surface ocean temperature to be estimated.

As mentioned in Chapter 8, in 1955 Cesare Emiliani published a large number of temperature curves based on twelve deep-sea cores, eight from the Swedish expedition and four from the Lamont core library.[23] Emiliani reported a 6°C change in the temperature of surface waters during the glacial interglacial cycles. He saw about fifteen temperature cycles spanning the entire Pleistocene. The temperature cycle minima showed reasonable correlation with solar radiation curves, supporting the Milankovitch theory, but a rigorous analysis was not attempted – that would only come much later. Several of the Lamont cores studied by Emiliani were also studied by the Ericson group at Lamont so that the qualitative curves based on foram assemblages could be compared with Emiliani's quantitative temperature curves. Overall, there was broad agreement between the two, but in detail there were important differences which led to substantial controversy: Was Emiliani's timescale, which was partly based on an assumed constant sedimentation rate, accurate? Did the apparent isotopic temperature actually represent ocean temperature or something else? If the isotopic composition of the ocean itself changed during glacial cycles then the foram isotopic temperatures would have to be corrected by an unknown amount.[8]

It was known that the continental ice sheets were isotopically light, that is enriched in ^{16}O, making the oceans isotopically heavier and displaying higher $^{18}O/^{16}O$ ratios during glacial periods. An English geochemist at Cambridge, Nicholas Shackleton, suggested that by comparing the isotopic composition of benthic (bottom-dwelling) forams with planktonic (shallow-dwelling) species in the same core, the question of the temperature significance of oxygen isotopes could be addressed. Using a core (named V28–238) raised by the Lamont research vessel *Vema* from the Pacific Ocean, paleomagnetic studies showed that the core penetrated the Brunhes-Matuyama geomagnetic reversal (Chapter 7) about 700,000 years ago, providing a critical anchor point for the second half of the Pleistocene.[13] A comparison of the oxygen isotope composition of benthic and planktonic forams showed that the patterns were almost identical. Since bottom ocean waters are close to 0°C and surface waters are much warmer, the similarity of the two curves suggests that temperature was not playing a major role, but rather that the isotopic composition of the ocean itself was controlling the patterns. If this was the case, then the isotopic patterns (e.g., those of Emiliani) were primarily recording the volume of ice on the continents. The isotopic curve from core V28 – 238 also agreed with available sea-level curves in several regions for that time period, confirming that sea level changes are related to continental ice volume (Figure 9.3).[13] This would make isotope stratigraphy even more significant – rather than indicating temperature alone, they now had significance in terms of global ice volume, a more important variable. Although Emiliani, in his pioneering study, had made a correction for a changing ocean isotopic composition, his correction was apparently too small; the actual surface ocean temperature variations were much smaller than his estimate (a maximum of 6°C).

An important international attempt to characterize Pleistocene climate was the Climate: Long-Range Investigation, Mapping, and Prediction Project (CLIMAP). Formed in 1971, it was funded by the National Science Foundation of the United States and other national organizations and headquartered at the Lamont-Doherty Geological Observatory. CLIMAP's first results were presented in a 1976 publication that showed the difference between surface temperature of the world's oceans for August at the Last Glacial Maximum, 18,000 years ago, and today.[24] Sea surface temperatures were based on foram and other microfossil assemblages from deep sea cores. Large ice sheets

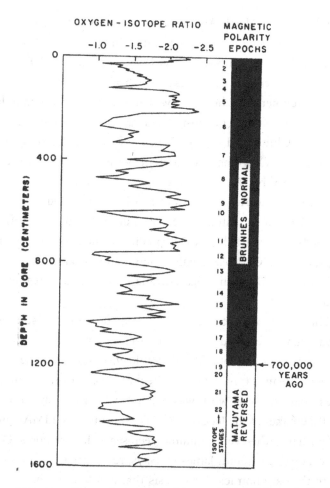

FIGURE 9.3 Oxygen isotope record for core V28–238 from the Pacific Ocean. The Brunhes/Matuyama magnetic reversal is shown, now known to be 780,000 years old. This was the first complete isotopic record of the late Pleistocene epoch. The isotopic variation is interpreted as indicating global ice volume rather than ocean temperature. More negative values of $\delta^{18}O$ indicate warm periods (odd-numbered stages).

Source: Shackleton, N. and Opdyke, N. 1973. Oxygen isotope and paleomagnetic stratigraphy of equatorial Pacific core V28–238: oxygen isotope temperatures and ice volumes on a 10^5 and 10^6 scale. *Quaternary Research,* v. 3, 39–55. Courtesy Elsevier.

were indicated for North America and northern Europe, but Alaska and large parts of Siberia were unglaciated. In the western Pacific, the Sea of Japan also saw temperature decreases of 8°C to 10°C. Equatorial Pacific saw temperature decreases as much as 6°C. The global average sea surface temperature, however, was only 2.3°C lower than today, a surprisingly low value. The most recent value by the Intergovernmental Panel on Climate Change (Table 5.2 of that publication) gives a value of 0.7–2.7°C, which is consistent, within error, with the CLIMAP value.[25] In the North Atlantic region, 18,000 years ago, the Gulf Stream crossed the Atlantic further south and traveled eastward to Iberia so that warm waters were not transported to high latitudes (Figure 9.4).[26] Remarkably, Croll predicted such a pattern in his 1864 paper. The CLIMAP results show a temperature decrease of 8–10°C (14.4–18°F) for the high-latitude north Atlantic region compared to today.

The theory of orbital variations as the driver of Pleistocene ice ages was finally resolved by an interdisciplinary group of scientists with expertise in microfossil paleontology, oxygen isotope stratigraphy, and advanced mathematical techniques. An important paper that resulted from this work was entitled "Variations in the Earth's Orbit: Pacemaker of the Ice Ages."[27] Workers on the CLIMAP project were familiar with several hundred deep-sea sediment cores. One of these workers at Lamont, James D. Hays, recognized that in order to evaluate the astronomical hypothesis that a core must have several properties, namely: it should be continuous for about 450,000 years to be statistically useful; it must have accumulated at a fast enough sedimentation rate so that bioturbation (biological disturbance of sediments) has not destroyed the short period signals; and it should be far from the influence of continental sediments. The only two cores to meet these criteria were from the southern Indian Ocean. Treating the core signals as a time series allowed application of sophisticated mathematical techniques (spectral analysis and band-pass filter analysis) to the isotopic signals they identified the three Milankovitch periods corresponding to precession of the seasons

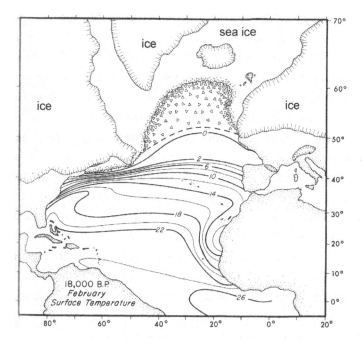

FIGURE 9.4 CLIMAP reconstruction of North Atlantic 18,000 years ago, showing winter surface sea temperatures (degrees Celsius). Note the closely spaced isotherms for the North Atlantic, indicating a more southern path for the Gulf Stream crossing the Atlantic to Europe at the latitude of Iberia. Triangles indicate calved icebergs.

Source: McIntyre, A. 1976. Glacial North Atlantic 18,000 years ago: A CLIMAP reconstruction. *Geological Society America Memoir* v. 145, 43–76. Courtesy Geological Society of America. Courtesy Cambridge University Press.

(23,000 years), axial tilt (41,000 years), and eccentricity (100,000 years). A hundred years after Croll addressed this problem, it was finally solved, but not until hundreds of deep sea sediment cores from all of the world's oceans were examined by scores of scientists with different specialities. The CLIMAP Project therefore played a fundamental role in solving this longstanding problem.

In addition to these three astronomical variations, it was also known that the composition of the Earth's atmosphere affected the Earth's surface temperature. John Tyndall, an Irish scientist, had shown by laboratory experiment in 1859 that carbon dioxide and methane gases blocked the passage of heat and light.[8] These are now known as greenhouse gases because they let in visible solar radiation through the atmosphere, but block the infrared radiation from going back out. Greenhouse gases therefore act as a one-way filter in the atmosphere and cause heating of the Earth's ground surface. The greenhouse effect of methane is twenty-five times more powerful than that of carbon dioxide, but because its concentration in the atmosphere is much less, it only produces about 30% of the warming compared to carbon dioxide. The Swedish scientist, Svante Arrhenius, later showed in 1908 that halving the carbon dioxide concentration in the Earth's atmosphere would result in a global temperature decrease of about 5°C (9°F), enough to cause an ice age. The most recent estimates by the Intergovernmental Panel on Climate Change gives a range of +1°C to +5°C for a doubling of CO_2 concentration, the large range being related to uncertainties in aerosols, the role of clouds, and continental use change (e.g., deforestation).[25] It is now thought that although greenhouse gases themselves do not initiate or end ice ages, their role is to amplify the effects produced by the orbital variations through various feedback mechanisms that are not yet fully understood.

ICE CORES

In order to study the role of greenhouse gases in climate change, researchers turned their attention in the 1990s to ice cores from polar regions. The National Ice Core Laboratory in Denver, Colorado, is operated by the United States Geological Survey and its purpose is to store ice cores drilled and extracted mainly from the polar regions. The facility is held at –35°C (–31°F), which is the optimum temperature to preserve the ice and the gases trapped in the ice. Visitors are given heavy parkas and thick mittens for a tour of the facility. The laboratory houses 17 kilometers (10.5 miles) of ice core from thirty-four drilling sites, largely from Greenland and Antarctica and other glacial regions. The ice cores themselves are about the same diameter as a large soda bottle – they are sliced in half lengthwise; one half is used for analyses, and the other half is stored for future study. Since the ice collections represent a great deal of effort and expense on the part of hundreds of technicians and scientists, there are three back up refrigeration systems at the Denver facility in case of a power failure.

When it snows in polar regions, each year the snow gets buried under more snow and eventually it turns into ice due to the overlying weight. Some of the air in the snow gets trapped in the ice as small bubbles. The pockets of trapped air represent samples of the atmosphere, and they record the chemical composition of the atmosphere at that time. The ice also records layers of ash from far away volcanic eruptions and dust blown in from arid regions. The annual layers in the ice can be individually counted, providing an accurate timescale. A complication to the simple counting of annual layers in the ice to provide a timescale is that there is some difference between the age of the ice and the age of the trapped gases – the gases tend to diffuse to deeper levels before pore spaces in the ice are sealed off, so that the gases are younger than the surrounding ice layers – in some cases by as much as a thousand years, depending on the depth in the core and the snow accumulation rate at that location. The age of the gases can be corrected, however, by using models based on gas diffusion. Results from four cores, two from Antarctica (Vostok and Dome C) and two from Greenland (GISP2 and GRIP) are now briefly described.

Collaboration between Russian, American, and French scientists resulted in the retrieval of a 3,623-meter-long core from the Vostok station in East Antarctica.[28] Drilling began in 1990 and was completed in 1998. The core covers four complete glacial cycles over a 420,000-year period, and each cycle shows a sawtooth pattern in which a rapid interglacial warming is followed by a slow prolonged glacial cooling (Figure 9.5).[29] The timescale for this core was calibrated by using two control points from the marine isotopic record (i.e., deep sea cores) at 110,000 years and 390,000 years. The intervening ages were interpolated by counting the number of annual ice layers. The solar radiation curve for June at 65° N latitude matches

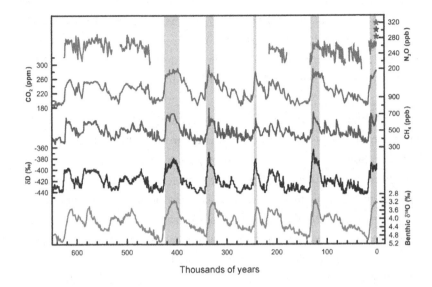

FIGURE 9.5 Variations in greenhouse gases CO_2 and CH_4 and hydrogen isotopes (δD) in the Vostok ice core, Antarctica. Parts of the N_2O curve are also shown (top curve). Also shown (bottom curve) is a compilation of oxygen isotopes ($δ^{18}O$) from deep-sea cores based on benthic microfossils (forams). Vertical shaded regions indicate interglacial warm periods. The figure shows that the greenhouse gases and hydrogen isotopes (a proxy for temperature) from the ice core are synchronous with each other and also synchronous with the oxygen isotope deep-sea record (a proxy for ice volume). A periodicity of about 100,000 years is apparent. Other periodicities of 41,000 and 23,000 years are also present, corresponding to astronomical variations of the Earth's orbit. Today's CO_2 concentration is about 400 parts per million.

Source: Jansen, E., Overpeck, J., Briffa, K. R., et al. 2007. Paleoclimate. In *Climate Change 2007; The Physical Science Basis. Contribution of Working Group I to the Fourth Assessment Report of the Intergovermental Panel on Climate Change.* Solomon, S., Qin, D., Manning, M., et al. (eds.). Cambridge University Press, Cambridge.

the ice core peaks very well using this timescale. The temporal resolution of the core is estimated to be ±5,000 years (for less than 110,000 years), and ±10,000 years for the remainder of the core. This relatively poor resolution makes it difficult to decide whether the greenhouse gases (CO_2 and CH_4) changes preceded or lagged behind the temperature changes, producing a cause-and-effect ambiguity (somewhat like the chicken-and-egg problem). To resolve this problem more precisely dated ice cores would be needed.

Chemical and physical analyses of the ice and the air trapped in the ice are of numerous types and include: isotopic analyses (δD_{ice} and $\delta^{18}O_{air}$), greenhouse gases CO_2 and CH_4 concentrations, and also dust levels in the ice. The deuterium analyses (δD_{ice}) are thought to reflect the air temperature at the drilling site, but more significantly, the oxygen isotope analyses ($\delta^{18}O_{air}$) are thought to reflect continental ice volume as noted already. Remarkably, the patterns of methane and carbon dioxide concentrations are almost identical to each other, and furthermore these patterns are very similar to the temperature proxy (δD_{ice}) (Figure 9.5). Even more remarkable, the marine record of oxygen isotopes of benthic forams from deep sea cores shows the same pattern (bottom curve, Figure 9.5), confirming that this record is a record of ice volume. Importantly, the maximum concentrations of both methane (660 parts per billion) and carbon dioxide (280–300 parts per million) over the 420,000 year period at Vostok are substantially lower compared to current values of these gases in the atmosphere today (1,800 parts per billion and 400 parts per million, respectively). This conclusion can be extended back 740,000 years to when the older EPICA (European Project for Ice Coring in Antarctica) Dome C core was examined.[30] The patterns in this core for the first four glacial cycles are very similar to those at Vostok, which is 560 kilometers away.

In central Greenland two cores were drilled almost simultaneously separated by only 28 kilometers by a European group (GRIP, the Greenland Ice Core Project) and an American group (GISP2, the Greenland Ice Sheet Project 2) during the period 1988 to 1993. The goal was to compare the records of the two ice cores to insure that the changes observed were real. The GRIP group reached bedrock first in 1992, followed by GISP2 in July of 1993, and both groups retrieved ice cores approximately 3 kilometers long. The GRIP (European) group published first,[31] while the GISP2 (American) hole was still being completed.[32] Their core penetrated the interglacial period prior to our present Holocene epoch, about 125,000 to 115,000 years ago, which is called the Emian interglacial. In contrast to the stable climate of the Holocene epoch, the GRIP Emian record showed a very unstable oxygen isotope record – an observation that, if correct, would have profound implications for our present climate and near future projections. In other words, stable interglacial climate could become suddenly unstable. This instability was not, however, observed in the Vostok record.

The GRIP group explained the instability as due to rapidly changing regimes in the North Atlantic Ocean currents.[31] This mistaken conclusion, however, was shortly corrected. The GISP2 group reached bedrock shortly thereafter and published a comparison of the GRIP[33] and GISP2[34] records in December of 1993. The GISP2 core did not display the instability seen in the GRIP core, which showed folding and deformation of the ice in their Emian section and meant the unstable record was an artifact of ice deformation and not due to climate change. Both coring groups concluded that one or both cores were disrupted at the Emian level and that the cores had no climate implications at that depth; the shallower sections (for times less than 110,000 years) of both cores were remarkably similar and showed major climate swings. The decision to drill two cores near one another, at great expense in terms of time and funding to both the European and American science foundations, proved to be a wise one – otherwise seriously false conclusions would have been drawn.[31]

The methane record for the Greenland GISP2 core for the last 40,000 years (which is very similar to GRIP) provides substantial insight into climate change during that interval, and is shown in Figure 9.6.[35] First, a note on terminology is in order. Within an ice age are glacial periods and interglacial periods. Within each glacial and interglacial cycle are much shorter cycles of cold and warm periods; these are called stadials and interstadials, respectively, to distinguish them from the longer glacial and interglacials. Concentrations of methane range from lows of about 400 parts per billion during cold periods (stadials) to highs of 750 parts per billion during

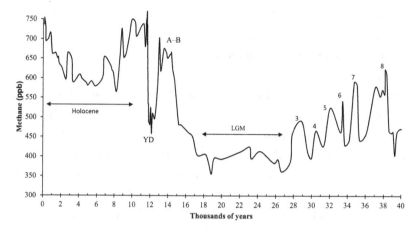

FIGURE 9.6 Methane concentrations (in parts per billion) from the GISP2 Greenland ice core for the past forty thousand years. The GRIP Greenland ice core shows a very similar record over this time period. High methane concentration reflect warm/wet climate and low concentrations reflect cold/dry climate. Numbers 3 through 8 refer to interstadials (warm periods) on a millennial time scale. LGM: Last Glacial Maximum; A–B: Allerød-Bølling warm period; YD: Younger Dryas cold period. The Holocene record shows a cold snap at 8,200 years ago. Today's atmospheric CH_4 concentration is about 1,800 parts per billion over twice the preindustrial value.

Source: Brook, E. J., Sowers, T., and Orchardo, J. 1996. Rapid variations in atmospheric methane concentration during the past 110,000 years. Science, v. 273, 1087–1091; and NOAA.

warm periods (interstadials). It can be seen that the Holocene epoch (0–11,500 years ago) is characterized by relatively high methane concentrations and the Last Glacial Maximum (17,000–26,000 years ago) is characterized by relatively low concentrations. The interval 28,000–38,000 years is characterized by several interstadials (numbered 3 through 8 in Figure 9.6) on a two-to-three-thousand-year cycle. This relatively short period is too short to be related to Milankovitch cycles.

The origin of these cycles can be gleaned from a study of North Atlantic deep-sea sediment cores. Normally, the sediment in these cores is fine silt and clays. Several horizons in these cores, however, show much coarser and distinctive horizons consisting of lithic fragments of basaltic glass, carbonates, and oxidized grain of sediment. These sediments are interpreted as ice-rafted sediments derived from continental rocks in the north Atlantic region. A very good correlation exists between these coarse sediment horizons[36] and cold periods (stadials) in the Greenland ice cores.[36,37] It is thought that calving off of large numbers of icebergs from the northern ice sheets caused the salinity (and density) of the North Atlantic Gulf Stream to decrease, reducing its strength in transporting warm water northward, causing cooling of the entire region (see Figure 9.4). It is not entirely clear what caused the icebergs to break off into the Atlantic at these periods – possibly a change in sea level. Following these cold conditions, abrupt warming occurred and caused the ice to retreat and the Gulf Stream to restart, returning warm water to higher latitudes. This interdisciplinary research, combining the study of continental ice cores and deep-sea sediment cores, provides clear evidence that the atmospheric greenhouse gases (CO_2 and CH_4), the oceans, and the ice sheets together act as a closely linked complex system, but the nature and the timing of these effects are still not yet fully understood.

Prior to the Industrial Revolution, the main source of methane in the atmosphere was wetlands, both tropical swamps and northern peat bogs such as those in Canada and Siberia. During the Last Glacial Maximum, these latter sources were probably frozen or covered by ice, leaving tropical wetlands as the main methane source during glacial and stadial cold periods. In wetlands, bacteria break down complex organic molecules to produce simpler molecules such as acetic acid (CH_4CO_2). In turn, the acetic acid is broken down to two greenhouse gases, namely CO_2 and CH_4. High temperatures tend to promote these reactions so that warm wet periods are expected to produce high methane concentrations and cold dry periods to produce low concentrations in the atmosphere. If these gases produced further warming, a positive feedback mechanism would be set up.

Alternatively, it has been suggested that the methane spikes in Figure 9.6 may have a marine source in the form of clathrates (methane molecules trapped inside ice molecules)[38] that become unstable on the continental margins due to warmer ocean temperatures.[39] More recent detailed isotopic studies of the methane in a Greenland ice core, however, do not support a marine source for this methane.[40] The methane troughs (stadials) in GISP2 are accompanied by peaks in dust concentration in the ice, indicating the atmosphere contained more dust during cold intervals. This dust corresponds to the wind-blown silt (loess), already mentioned, and indicates the glacial climate (stadials) was windy and dry. In contrast, the interstadials were temperate and wet.[41]

A sharp rise in methane occurred 14,700 years ago at the end of the Last Glacial Maximum (Figure 9.6) due to melting of continental ice, which also produced a sharp rise in global sea level. The geologic record of corals on Tahiti in the Pacific show a rise in sea level of twenty meters in the interval 14,700–14,300 years ago (corresponding to a 0.5 meter rise per decade). This melting is thought to have been caused by melting in the Antarctic ice sheet resulting in changes to the North Atlantic deep water current, leading to the Bølling warming event (A-B in Figure 9.6).[41] The rise in sea level would have increased the area of coastal wetlands and increased methane emissions.

This warming had important implications for early European humans. During the Last Glacial Maximum (LGM) archaeological evidence indicates the human population was largely confined to refuges in southern Europe, but that a large northward migration took place at the time of the A-B warming.[42] Much of the Paleolithic cave art in France and northern Spain is attributed to this time period.[43] The warming lasted only 1,500 years before the cold suddenly returned in the Younger Dryas (YD, Figure 9.6). This cold period lasted 1,200 years before the climate warmed again, marking the beginning of our current Holocene epoch. The YD cooling is attributed to the same cause as the earlier cyclic cooling, namely calving of large amounts of ice off the northern continental ice sheets into the Atlantic Ocean resulting in a shutdown of the Gulf Stream. An alternative explanation of the YD cooling is that a glacial lake (Lake Agassiz) in the vicinity of the present-day Great Lakes drained into the north Atlantic by way of the St. Lawrence seaway, again causing the Gulf Stream current to shut down.[44]

REFERENCES

1. DeBeer, G. 2008. Ignatz Venetz. *Complete Dictionary of Scientific Biography*, v. 13, 604–605. Charles Scribner's Sons, Detroit.

2. DeBeer, G. 2008. De Charpentier, J. *Complete Dictionary of Scientific Biography*, v. 3, 210–211. Charles Scribner's Sons, Detroit.

3. De Charpentier, J. 1835. Notice sur la cause probable du transport des bloc erratiques du basin du Rhône. *Annales des Mines*, v. 8, 219–236.

4. De Charpentier, J. 1841. *Essai sur les glacier et sur le terrain erratiques du basin du Rhône*. Privately published, Lausanne.

5. Agassiz, L. 1837. Des glaciers, des moraines et des blocs erratique: Discours pronounce à l'ouverture des séances de la Société Helvétique des Sciences Naturalles. *à Neuchâtel*, v. 12, 369–393.

6. Rudwick, M. 2008. Snowball Earth? (1835–1840). In *Worlds before Adam*. University of Chicago Press, Chicago.

7. Lyell, C. 1830–1833 (1997). *Principles of Geology*. Penguin Classics, London.

8. Imbrie, J. and Imbrie, K. P. 1986. *Ice Ages: Solving the Mystery*. Harvard University Press, Cambridge, MA.

9. Agassiz, L. 1841. Glaciers and the evidence of their having once existed in Scotland, Ireland and England. *Proceedings of Geological Society of London*, v. 3, 327–332.

10. Buckland, W. 1841. Evidences of glaciers in Scotland and the north of England. *Proceedings of Geological Society of London*, v. 3, 332–336.

11. Maclaren, C. 1842. The glacial theory of professor Agassiz of Neuchâtel. *American Journal of Science*, v. 42, 346–365.

12. Whittlesey, C. 1868. Depression of the ocean during the ice period. *American Association for the Advancement of Science*, v. 16, 92–97.

13. Shackleton, N. and Opdyke, N. 1973. Oxygen isotope and paleomagnetic stratigraphy of equatorial Pacific core V28-238: oxygen isotope temperatures and ice volumes on a 10^5 and 10^6 scale. *Quaternary Research*, v. 3, 39–55.

14. Potter, B. A., Holmes, C. E., and Yesner, D. R., 2013. Technology and economy among the earliest prehistoric foragers in interior eastern Beringia. In *Paleoamerican Odyssey*, 81–103. Graf, K. E., Ketron, C. V., and Waters, M. R. (eds.). Texas A&M Press, College Station.

15. Wright, G. F. 1890. *The Ice Age in North America: And its bearings upon the antiquity of man.* Appleton & Co., New York.

16. Gilbert, G. K. 1890. *Lake Bonneville.* U. S. Geological Survey Monograph 1, Washington, DC.

17. Croll, J. 1864. On the physical cause of the change of climate during geological epochs. *Philosophical Magazine*, v. 28, 121–137.

18. Croll, J. 1875. *Climate and Time.* Appleton & Co., New York.

19. Milankovitch, M. 1920. *Théorie mathématique des phénoménes thermique produits par la radiation solaire.* Gauthier-Villars, Paris.

20. Imbrie, J. 1982. Astronomical theory of the Pleistocene ice ages: A brief historical review. *ICARUS*, v. 50, 408–422.

21. Bassard, D. C. 2004. Website for index of H. M. S. Challenger scientific reports: www.thcenturyscience.org/HMS-INDEX/-index.htm. Accessed June 12, 2016.

22. Ericson, D. B., Broecker, W. S., Kulp, J. L. and Wollin, G. 1956. Late-Pleistocene climates and deep-sea sediments. *Science*, v. 124, 385–578.

23. Emiliani, C. 1955. Pleistocene temperatures. *Journal of Geology*, v. 63, 538–578.

24. CLIMAP members. 1976. The surface temperatures of the Ice-Age Earth. *Science*, v. 191, 1131–1138.

25. Stocker, T. F. et al. (eds.). 2013. *Climate Change 2013: The Physical Basis.* Intergovernmental Panel on Climate Change. Cambridge University Press, Cambridge.

26. McIntyre, A. 1976. Glacial North Atlantic 18,000 years ago: A CLIMAP reconstruction. *Geological Society of America Memoirs*, v. 145, 43–76.

27. Hays, J. D., Imbrie, J., and Shackleton, N. 1976. Variations in the Earth's orbit: Pacemaker of the Ice Ages. *Science*, v. 194, 1121–1128.

28. Petit, J. R., Jouzel, J. Raynaud, D., et al. 1999. Climate and atmospheric history of the past 420,000 years from the Vostok ice core, Antarctica. *Nature*, v. 399, 429–436.

29. Jansen, E., Overpeck, J., Briffa, K. R., et al. 2007. *Paleoclimate In Climate Change 2007: The Physical Science Basis.* Contribution of Working Group I to the Fourth Assessment Report of the Intergovernmental Panel on Climate Change. Solomon, S., Qin, D., Manning, M., et al. (eds.). Cambridge University Press, Cambridge.

30. EIPCA. 2004. Eight glacial cycles from an Antarctica ice core. *Nature*, v. 429, 623–628.

31. Dansgaard, W., Johnsen, S. J., Clausen, H. B., et al. 1993. Evidence for general instability of the past climate from a 250-kyr ice-core record. *Nature*, v. 364, 218–220.

32. Greenland Ice Core Project (GRIP) members. 1993. Climate instability during the last interglacial recorded in the GRIP ice core. *Nature*, v. 364, 203–207.

33. Taylor, K. C., Hammer, C. U., Alley, R. B., et al. 1993. Electrical conductivity measurements from the GISP2 and GRIP Greenland ice cores. *Nature*, v. 366, 549–552.

34. Grootes, P. M., Stuiver, M., White, J. W. C., Johnsen, S., and Jouzel, J. 1993. Comparison of oxygen isotope records from the GISP2 and GRIP Greenland ice cores. *Nature*, 366, 552–554.

35. Brook, E. J., Sowers, T., and Orchardo, J. 1996. Rapid variations in atmospheric methane concentration during the past 110,000 years. *Science*, v. 273, 1087–1091.

36. Bond, G. and Lotti, R. 1995. Iceberg discharges into the North Atlantic on millennial time scales during the last glaciation. *Science*, v. 267, 1005–1010.

37. Bond, G., Broecker, W., Johnsen, S., et al. 1993. Correlations between climate records from North Atlantic sediments and Greenland ice. *Nature*, v. 365, 143–147.

38. Nisbet, E. G. 1990. The end of the Ice Age. *Canadian Journal of Earth Science*, v. 27, 148–157.

39. Kennett, J. P., Cannariato, K. G., Hendy, I. L., and Behl, R. J. 2003. *Methane hydrates in Quaternary climate change: The clathrate gun hypothesis*. American Geophysical Union Monograph, Washington, DC.

40. Bock, M., Schmitt, J., Möller, L., et al. 2010. Hydrogen isotopes preclude marine hydrate CH_4 emissions at the onset of Dansgaard-Oeschger events. *Science*, v. 328, 1686–1689.

41. Weaver, A. J., Saenko, O. A., Clark, P. U., and Mitrovica, J. X. 2003. Melt pulse 1A from Antarctica as a trigger of the Bølling-Allerød warm interval. *Science*, v. 299, 1709–1713.

42. Gamble, C., Davies, W., Pettitt, P., and Richards, M. 2004. Climate change and evolving diversity in Europe during the last glacial. *Philosophical Transactions of Royal Society of London, B*, v. 359, 243–254.

43. O'Hara, K. D. 2014. *Cave Art and Climate Change*. Archway Publishing, Bloomington, IN.

44. Broecker, W. S., Kennett, J. P., Flower, B. P., et al. 1989. Routing of melt-water from the Laurentide ice sheet during the Younger Dryas cold episode. *Nature*, v. 341, 318–321.

10 Geology and Evolution of the Moon

At length, by sparing neither labor or expense I succeeded in constructing for myself an instrument so superior that objects seen through it appeared thirty times nearer ... I betook myself to observations of the heavenly bodies; and first of all, I viewed the Moon.

– Galileo Galilei, 1610[1]

Galileo Galilei (1564–1642) appears to have been the first to observe the moon through a telescope.[1] He focused on the boundary between the region of the moon in shadow and in sunlight as the sun rose on the lunar landscape (called the terminator by astronomers). He saw the boundary was highly irregular and correctly inferred that the topography of the moon "was neither smooth nor uniform" and also that the circular "spots" he observed on a variety of scales were depressions (craters) rather than protuberances.[1] He also noted that bright spots in the shade were topographically elevated regions, and he estimated their height. Galileo was put under house arrest by the Vatican and remained there for the rest of his life because of his views on his new cosmology regarding the motion of the planets.

The near side of the moon always faces toward the Earth because it is gravitationally locked so that it rotates once for every orbit of the Earth. A basic observation that anyone could have made prior to Galelio's work is that the nearside of the moon consists of two types of terrain, dark smooth areas in contrast to highly cratered brighter regions, long known as the *maria* (Latin for seas) and the highlands, respectively. In keeping with the Latin terminology, the highlands are also known as *terra* or *terrae* in much of the modern literature. The maria were suspected by many as consisting of basalt, but it wasn't until samples were returned to Earth by the Apollo program that this was confirmed.[2] Prior to the Apollo program, many investigators favored a volcanic origin for the moon's craters and it was Gilbert, the director of the United States Geological Survey who was best known for his work on Lake Bonneville glacial deposits (Chapter 9), who favored a meteorite impact origin in his classic 1893 paper.[3] Others favoring an impact origin were Baldwin in 1949[2] and Urey in 1951.[4]

LUNAR STRATIGRAPHY

The first investigators to apply traditional stratigraphic principles, such as superposition and cross-cutting relationships, to the lunar surface were United States Geologic Survey geologists Eugene M. Shoemaker and Robert J. Hackman.[5] Their study was prepared on behalf of the National Atmospheric and Space Administration (NASA) in anticipation of the imminent exploration of the moon. Using ground-based telescope photographs at a scale of one to a million, their objective was to establish a relative lunar timescale. They focused their attention initially on a relatively small quadrangle close to the Copernicus crater, which is near the southern margin of Mare Imbrium (Figure 10.1). By looking at the temporal relationships between crater ejecta, crater infill, smooth and rough formations, and areas of different albedo (reflectivity), they produced the stratigraphic column shown in Table 10.1 – which has been only slightly modified since their pioneering study. The subsequent study of Don F. Wilhelms and John F. McCauley in 1971 produced a colored map of the entire nearside of the moon at a scale of one to five million based on the statigraphy shown in Table 10.1.[6] They divided the Imbrian system into upper and lower series to replace the earlier Procellarian period of Shoemaker and Hackman, which largely corresponded to the maria basalt outpourings. Wilhelms and McCauley argued that the basaltic maria were not a time stratigraphic unit because they were of different age in different basins, and so replaced it with a rock stratigraphic unit: the Upper Imbrian.[6] They also showed the Frau Mauro formation as the ejecta blanket of the

Table 10.1 *Lunar stratigraphy*

Shoemaker & Hackman[5] (1962)	Wilhelms & McCauley[6] (1971)	Wilhelms[7] (1987)
Copernican	Copernican	Copernican
Eratosthenian	Eratosthenian	Eratosthenian (Maria)
Procellarian (Maria)	Upper Imbrian (Maria)	Upper Imbrian (Maria)
Imbrian	Lower Imbrian	Lower Imbrian
Pre-Imbrian	Pre-Imbrian	Nectarian
		Pre-Nectarian

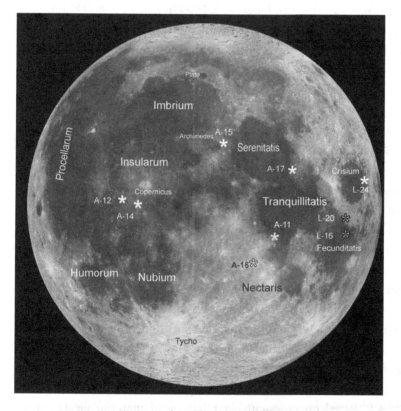

FIGURE 10.1 Location of Apollo and Luna (L16, L20, and L24) landing sites. Maria are indicated, and the craters Plato (97 kilometers across), Archimedes (80 kilometers), and Tycho (90 kilometers) are also indicated. *Sources:* NASA and Lick Observatory. Courtesy Regents of the University of California.

Imbrium basin on their new map. Subsequently, the ejecta blanket of the older Nectaris basin was recognized, called the Janssen formation, and the Pre-Imbrian system was subsequently replaced with Nectarian and Pre-Nectarian systems (Table 10.1).[7]

MULTI-RINGED BASINS

The largest features on the moon, observable with the naked eye, are the large circular basins, now filled (or partially filled) with smooth basalt lava flows – the maria referred to above.[8] The basins are usually enclosed by three or more rings representing uplifted scarps. The innermost ring can range from 250 kilometers to 600 kilometers across.[6] These basins were excavated during the lunar heavy bombardment period, a time when asteroid collisions were common in the solar system about 3.9×10^9 years ago (3.9 Ga).[9] These impact basins are absent on Earth because of vigorous tectonic activity, but they are well-preserved on the moon. The origin of these features was not understood until the Orientale multi-ring basin on the far side of the Moon was observed by orbiting spacecraft (Figure 10.2). This is the youngest and least eroded of these basins and only the innermost ring is partly filled with mare basalt, so its overall structure is revealed. Orientale is the key to understanding the multi-ring basins on the nearside. The diameter of the structure is about 900 kilometers based on the outer ring (Montes Cordillera), and an asteroid about 50 kilometers in diameter is estimated to have caused this structure. By comparison, the asteroid that is inferred to have caused the end Cretaceous extinctions on Earth is estimated to have been ~10 kilometers across (Chapter 2). Although the origin of multi-ring structures is still debated, the rings are thought to represent shockwave fronts frozen into the target rock at the time of impact. Other interpretations argue that the rings are later subsidence structures.[10] Based on cross-cutting relations of material ejected from the impact basins and the overlap of their outer margins, the following relative sequence from youngest to oldest is indicated (see Figure 10.1)[10]:

FIGURE 10.2 Photograph of the Mare Orientale multi-ringed basin (900 kilometers across). Note mare basalt flooding the central ring. *Source:* NASA (Orbiter 4).

Orientale→ Imbrium→ Crisum→ Nectaris→ Humorum→ Serenitatitis→ Tranquillitatis→ Fecunditatis. Most of these basins are in the relatively narrow age range of 3.85–3.95 Ga,[9]

Some of the stratigraphic evidence that the mare basalt infilling of the basins came later and is not directly related to the formation of the basins themselves is summarized as follows[6]:

- The maria volcanics truncate all ejected basin deposits.
- The volcanics appear to have a narrower age range than the impact basins themselves, based on the number of superposed craters, which show various states of preservation.
- Post-impact basins but pre-mare craters (for example, the Archimedes crater within the Imbrium basin) are filled with basalt, indicating an interval of time passed between impact basin and mare infill.

The structural level of mare infill varies from basin to basin – Nectaris is filled to the first ring, Crisium and Imbrium to the second and third, respectively, suggesting an internal origin for the fill rather than an impact origin; the depth of fill is estimated to be between 0.5 kilometers and 4 kilometers.[7] That the mare basalts escaped the cataclysmic bombardment at 3.9 Ga indicates they are younger than this event.

THE APOLLO PROGRAM

In 1961, John F. Kennedy (1917–1963), speaking before the US Congress, set the goal of landing a man on the moon and returning him safely to Earth by the end of that decade. The political backdrop to this speech was the Cold War, when in 1957 the Soviet Union put a man into Earth's orbit aboard Sputnik I, marking the beginning of the space age. Kennedy recognized in his speech that the United States was behind in manned space exploration. His goal was achieved when NASA lofted Neil Armstrong onto the moon aboard Apollo 11 in 1969, meeting the short decade-end deadline.

The Apollo program sent six successful manned missions to the moon between 1969 and 1972 (Apollo 11, 12, 14, 15, 16, and 17; Apollo 13 was aborted). Table 10.2 shows that the duration of extravehicular activity (EVA), the distance traversed by the astronauts on the surface, and the weight of rock samples returned increased with each mission, indicating increasing technological sophistication and confidence of NASA over a three-year period. (Parenthetically, the unmanned Russian Luna missions, L16, L20, and L24, returned gram quantities of lunar soil.) The primary research results of lunar

Table 10.2 *The Apollo missions*[42]

Mission	Landing site	EVA (hours)	Distance traversed (kilometers)	Rocks returned (kilograms)
11	Mare Tranquillitatis	2.24	–	21.7
12	Mare Insularium	7.6	1.35	34.4
14	Fra Mauro	9.2	3.45	42.9
15	Hadley-Apennines	18.3	27.9	76.8
16	Descartes	20.1	27.0	94.7
17	Taurus-Littrow	22	30.0	110.5

scientists are published in the Proceedings of the Lunar Science Conference held in Houston, Texas, each year from 1970 to 1992, totaling 45 volumes, each approximately two thousand pages per volume. For this chapter, the author consulted volumes corresponding to the second, fourth, fifth, and twelfth Lunar Science Conferences, which include research from all of the lunar missions.

All of the Apollo landing sites lie within or close to the boundary of the maria areas, with the exception of Apollo 16, which landed entirely within a southern highland region (Figure 10.1). A brief summary of the results of each Apollo mission is given based on the lunar sample preliminary examination teams (LSPET reports) and also on the geology of the sites published by *Science* magazine after each mission. In the first three missions, due to fear of spreading pathogens, returned samples were quarantined for several weeks. In addition it was unknown whether the Earth's oxygen-rich atmosphere would degrade the samples. Preliminary examination of the rocks was therefore done through space-suit arms in a cabinet under vacuum.[11] Specially designed equipment was also used for rock analysis under quarantine conditions. The building to which the samples were returned in Houston, Texas, also had biological barriers installed. After the third mission it was concluded there was no danger of biological contamination and no degradation of the samples was observed. Initially the astronauts themselves were also quarantined for three weeks.

Before discussing the composition of the returned lunar samples, a brief review of some mafic igneous rock terminology may be in order. Gabbro is the coarse-grained (plutonic) equivalent of basalt and consists of two main minerals: calcic plagioclase (a Na-Ca solid solution feldspar) and Ca-rich pyroxene (a Ca-Fe-Mg solid solution). Norite is a more mafic variety of gabbro with calcic plagioclase and Ca-poor pyroxene (a Fe-Mg solid solution). With an increase in plagioclase content, gabbros and norites pass into a largely feldspathic rock: anorthositic gabbro or anorthositic norite. Pure plagioclase feldspar produces anorthosite, a monomineralic rock; dunite is also a monomineralic rock composed of olivine. Troctolite is an olivine-bearing anorthosite. With an increase in pyroxene and olivine, gabbro and norite pass into ultramafic rocks such as peridotite. There are rare examples of each of these rock types, mainly as small clasts in breccias in the Apollo sample collections.

The Apollo 11 lunar module landed in the southwestern part of Mare Tranquillitatis on July 20, 1969 (Figure 10.1). Immediately surrounding the landing site were several small (tens of meters across) shallow craters.[11] Armstrong first collected contingency samples near the Lander in case the mission was aborted early, so that at least some samples would be returned. He and astronaut Edwin "Buzz" Aldrin collected a total of 22 kilograms of rock over a 160-minute EVA. Half of this rock consisted of rocks greater than 1 centimeter in size and half of finer materials. Cores of the lunar soil were also taken by Aldrin.

The rocks collected consisted of fine-grained vesicular volcanic rock and medium-grained vuggy volcanic rock and breccias. The major minerals present in the volcanic rocks and the breccias were Ca-rich pyroxene (50%), plagioclase (30%), opaque minerals (20%), and olivine (0–5%). In discussing the mineralogy of lunar rocks, it is important to bear in mind the two main types of pyroxenes: calcium-rich pyroxenes (also called clinopyroxenes, or cpx) and calcium-poor pyroxenes (also called orthopyroxenes, or opx). These distinctions become important in discussions of the petrogenesis of the lunar highlands (which are commonly calcium rich) and the maria rocks (which are relatively calcium poor). No hydrous phases were observed (which holds true for all Apollo mission samples). Glass was present in the matrix of many rock samples and spheres of glass were common in the soil, also a feature common to all subsequent mission sites. The glass is probably either impact melts or products of volcanism.[10] Most rock samples showed evidence of shock or impact metamorphism (e.g., fractured and crushed minerals and glass). The breccias had a complex history as some breccia clasts were themselves brecciated (so-called polymict breccias). The preliminary results of the findings of Apollo 11 were presented in a special moon-related issue of *Science* magazine in 1970, which includes a summary of the Apollo 11 Lunar Science Conference held in Houston, Texas.[12]

The chemistry of the fine-grained and medium-grained volcanic rocks and the breccias were similar to each other, but different from terrestrial rocks and meteorites. The lunar rocks were unusually high in Ti, Zr, Y, and Cr, and unusually low in Na and K compared to terrestrial rocks; as noted already, no hydrous phases were present. An average major element chemical composition for Apollo 11 basalt samples (in weight percent) is: SiO_2 (40); Al_2O_3 (10); TiO_2 (10); FeO (20); MgO (8); CaO (10). Note the high FeO and TiO_2 contents. Preliminary analysis by the K-Ar method showed the Apollo 11 rocks to be very old: between 3×10^9 and 4×10^9 years (4–3 Ga).

The Apollo 12 and nearby Apollo 14 landing sites were southwest of Copernicus crater in Mare Insularum (Figure 10.1). Apollo 12 samples were mare basalts, but also included olivine norites (Ca-poor pyroxene and plagioclase) and troctolites (plagioclase and olivine), and olivine was common in basalts in contrast to those at the Apollo 11 site.[13] Impact melt glass was also common. The basalts were lower in Ti compared with Apollo 11 basalt. Many of the samples define a smooth sequence with decreasing MgO content and other elements, suggesting crystal fractionation (of olivine and or pyroxene), as is commonly seen in terrestrial basaltic rocks. One sample displayed a very distinctive chemistry with high incompatible element concentrations (elements that do not easily fit into common mineral structures). This rock component also occurs at the landing sites of Apollo 14, 15, 16, and 17, as well as in Luna 24 soil samples. These samples are now referred to as KREEP-rich (rich in potassium, rare earth elements, and phosphorous). In the petrology of lunar rocks, this enigmatic rock *component* (a pure KREEP rock has not yet been found) has a vast literature because of its unique composition.[14,15] Apollo 12 basalts produced the youngest mare rock with K-Ar ages of 1.7×10^9 to 2.7×10^9 years (2.7–1.7 Ga). Apollo 14 landed nearby on the Frau Mauro Formation which, as noted already, is thought to be the pre-mare ejecta blanket from the Imbrium basin to the north. Most of the samples were impact-melt breccias, as might be expected at a site near a major impact basin.[16]

Apollo 15 was an ambitious mission with three EVAs allowing sampling of three different areas.[17] The landing site was on mare material on the eastern edge of Mare Imbrium near the frontal scarp of the Appennine Mountains, which form the outer uplifted ring of the Imbrium basin. The only rocks seen in outcrop by any of the Apollo missions occur in Hadley Rille, a sinuous volcanic channel that runs through the landing site. Nearby (about 500 meters) to the northwest are three closely spaced craters: Archimedes (80 kilometers across) lies within the Imbrium basin, but itself is filled with mare material and is therefore pre-Mare in age (Upper Imbrian; Table 10.1). The other two craters, Autolycus (39 kilometers) and Aristillus (56 kilometers), both without mare fill, are Copernican in age (Table 10.1), and rays from these craters cross the landing site. The mission sampled the Appennine Front, Hadley Rille, and mare basalts.[18] The Appennine Front rocks included meta-igneous rocks such as noritic anorthosite and troctolitic anorthosite thought to represent pre-mare highland crust. Also a green glass was found in the soil with a very primitive composition (high MgO) and an age of 3.3×10^9 years (3.3 Ga), suggesting it is related to the mare basalts.[18]

Apollo 16 is the only landing site well outside mare basins (Figure 10.1).[19] One of the goals was to date the Nectaris basin by sampling the ejecta sheet with a view to sampling the deeper lunar crust. The rocks sampled were all high in CaO and Al_2O_3 and rich in plagioclase, and corresponded to the plutonic rock anorthosite – although they are all highly crushed (i.e., cataclastic). Together with remote sensing data from lunar orbiters, this mission indicated the lunar crust is rich in anorthosite. This site is also rich in KREEP components. The Nectaris basin was dated to about 4.2 Ga by the ^{40}Ar-^{39}Ar method on breccias,[20] but more recent reviews place it as younger at 3.9 Ga.[9]

The last of the Apollo missions, Apollo 17 (in December 1972), was unique in that one of its four crew members, Harrison (Jack) Schmitt, was a geologist (Ph.D., Harvard, 1964). Astronauts Eugene Cernan (1934–2017) and Schmitt traversed about 30 kilometers on the lunar surface over a three-day period, and returned about 110 kilograms of rock and soil. The Apollo 17 landing site was in a valley (a graben) between the second and third outer rings of Mare Serenititus in the Taurus Mountains-Littrow region (Figure 10.1).[21] The landing site graben is floored by basalt and surrounded by uplifted massifs to the north and south thought to represent pre-mare "basement." The soil (or regolith) was 15 meters thick at this site compared to 3 meters at the Apollo 11 site. The samples returned are amongst the most variable of any mission.[22] Samples included basalts, several types of breccia including brecciated dunite and brecciated norite (a Mg-rich gabbro), and glass-bonded agglutinates (minerals and lithic fragments bonded by brown glass), as well as unusual orange glass droplets in the soil. Clasts in some breccias consisted of coarse-grained anorthosite and anorthositic norite, thought to represent primordial lunar crust.

The basalts are all high-titanium basalts similar to those sampled at the Apollo 11 site. The massif rocks consist of brecciated anorthositic gabbros. Many of these breccia clasts have KREEP-like trace element characteristics. The soils appear to be a mixture of mare basalt and anorthositic gabbro. The unusual orange glass in the soil is very high in titanium (about 10 weight percent), and though similar glass compositions have not previously been identified, such magmas are potentially parental to high-titanium mare basalts found at the Apollo 11 and 17 sites.

THE HIGHLANDS AND THE MAGMA OCEAN HYPOTHESIS

Terrestrial igneous petrologists learned an enormous amount about mafic igneous processes when they studied layered mafic and ultra-mafic intrusions on Earth, such as the Skaergaard intrusion in Green-land or the Bushveld intrusion in South Africa. In these settings the rocks had a simple stratigraphic relationship to each other, the rocks were largely fresh and igneous textures were well preserved, and even the original magma composition was sometimes preserved in chilled margin samples.

The sample sets returned by the Apollo missions presented an entirely different and more challenging set of problems. The lunar-cratering process (in particular the cataclysmic bombardment at around 3.9 Ga) pulverized the crustal layering so that the depth of excavation in the crust of the returned samples is largely unknown. Moreover, shock metamorphic effects (brecciation, cataclasis, and melting) largely destroyed the original igneous textures. One researcher, Paul Warren, entitled a series of his papers, "The search for nonmare pristine rocks, the 3rd foray ... the 7th foray...the 8th foray,"[23] which hints at a certain amount of exasperation on the part of that author. One small pristine clast (consisting of norite) from an Apollo 15 breccia did, however, yield a Sm-Nd internal isochron age of 4,460 ± 70 Ma (4.46 Ga), which appears to record the age of the lunar crust, if not the moon itself (Figure 10.3).[24] This age is also close to the age of the Earth of 4.55 Ga (Chapter 3). In this regard, two samples from the Apollo 11 collection plot on the same Pb-Pb isochron shown in Chapter 3.[25] Several pristine rock samples from Apollo 17 yield only slightly younger ages.[23] What was the composition of this lunar crust?

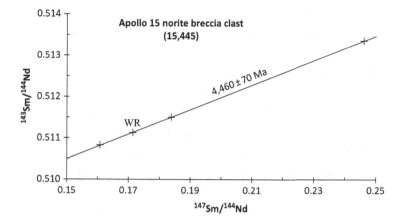

FIGURE 10.3 Samarium-neodynium internal isochron for pristine breccias clast (15,445) from Apollo 15 indicating an age of 4.46 × 10⁹ years (4.46 Ga). WR: whole rock; +: different mineral density fractions.
Source: Shih, C. Y., Nyquist, L. E., Dasch, E. J., et al. 1993. Ages of pristine norite clasts from lunar breccias 15445 and 15455. Geochimica et Cosmochimica Acta, v. 57, 915–931. Courtesy Wiley.

In a study of the Apollo 11 sample collection, a group of researchers identified shock metamorphosed anorthosites and gabbroic anorthosites in 4% of the samples they examined.[26,27] Together with the relatively simple chemical analysis provided by the pre-Apollo Surveyor 7 unmanned vehicle that landed near the Tycho crater in the southern highlands in 1969 (Figure 10.1),[28] these authors concluded that the lunar crust was largely composed of anorthosite. Furthermore they proposed that this crust of plagioclase (25 kilometers thick with a density of 2.9 g/cc) floated above a substratum of gabbro (3.0 g/cc when liquid), when a large fraction of the moon was molten shortly after its accretion.[26,27] Although they did not use the term, this has become to be known as the Magma Ocean hypothesis.[23] It is interesting to note that the most robust and widely accepted origin for the lunar crust came from a study of just 4% of the first Apollo mission samples. Subsequent missions identified a Mg-rich group of rocks, mainly norites, troctolites, several gabbros, and a dunite. On a diagram of Ca versus Mg, the anorthosites and Mg-rich group, however, plot in distinct regions and appear to be unrelated to each other.[23] They may represent intrusions from deeper Mg-rich cumulates into the anorthositic crust above.

Subsequent missions after Apollo 11 also identified a KREEP component (rich in potassium, rare earth elements, and phosphorous) in highland rocks (mainly in highland breccias and soils). This component also has very high abundances of incompatible elements such as Cs, Rb, Ba, Th, U, Zr, Hf and Nb. These elements do not fit into the major rock-forming mineral phases and remain in the liquid during cooling. Subsequent models of the magma ocean hypothesis put its depth at between tens and hundreds of kilometers. As the magma cooled it first crystallized olivine and pyroxene, which sank to the bottom of the magma ocean to form cumulates. This eventually (after ~70% crystallization) led to crystallization of plagioclase as calcium increased in the magma. The plagioclase was inferred to have floated to form the crust; it was subsequently shown experimentally that plagioclase (density 2.9 g/cc) would indeed float in an anhydrous lunar magma (density 3.0 g/cc).[29] The KREEP component is best explained as the last dregs of magma (1% or 2% liquid) as the magma ocean neared total crystallization. The high concentrations of the rare earth elements in KREEP (up to a thousand times that in the crust) requires a very large magma ocean to supply these elements in sufficient quantity.[30] Subsequent meteorite bombardment pulverized the crust and mixed the KREEP component into polymict breccias. The major element composition of KREEP is unknown since a KREEP rock has not been identified in the lunar collection.

During the Apollo 15 and 16 missions, while the astronauts were collecting samples on the surface, the orbiting command modules performed two remote sensing experiments – one with an X-ray spectrometer, the other with a gamma-ray spectrometer.[31] The gamma ray spectrometer provided data on the concentrations of radioactive nuclides such as K, U, and Th in the lunar crust and the maria. These elements acted as heat sources and are important in understanding the thermal evolution of the moon. The X-ray spectrometer provided data on Si, Al, and Mg concentrations for both the near side and the far side. These data are largely consistent with the magma ocean hypothesis in which the highlands showed higher Al/Si ratios compared to the maria (anorthosites having a higher Al/Si ratio compared to basalts).

A surprising finding in the lunar rock samples was the very strong coherence between some elements, stronger than seen in terrestrial rocks. Based on the periodic table of elements, we expect elements in the same vertical group to be geochemically coherent (e.g., K and Rb, or Zr and Hf), but due to some hydrothermal processes on Earth that are absent on the moon, the coherence is not always perfect. For lunar rocks, however, if the concentration of one element is known, the concentration of a geochemically coherent element can be accurately predicted. Using the sparse data from the remote-sensing experiments on Apollo 15 and 16 and chemical analyses of the rocks themselves, investigators attempted to reconstruct the average major and trace element composition of the moon and its crust.[10,32–34] Unfortunately the agreement between these estimates was not particularly good.[35] The Australian geochemist, A. E. Ringwood, based on his high-pressure experimental work using lunar bulk compositions, maintains that after the feldspathic lunar crust separated, the mafic residue was not capable of producing mare basaltic compositions by remelting. Ringwood proposed his own model for the origin of mare basalts by the melting of the undifferentiated lunar interior at very deep levels.[35] This model however does not explain the complementary nature of the rare earth elements patterns of the crust and mare basalts.

MARE BASALTS

The history of lunar volcanism was reviewed in time and space by the planetary geologist James Head in 1976.[36] Figure 10.4 summarizes the published ages of mare basalts up until 1980, and shows the ages range from 3.8 Ga to 3.1 Ga, indicating that they post-date the cataclysmic bombardment of circa 3.9 Ga that formed the multi-ringed basins.[37]

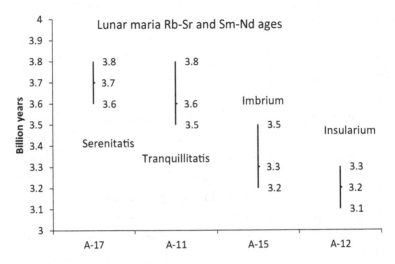

FIGURE 10.4 Rb-Sr and Sm-Nd ages for lunar maria. The mean age and range in age for each basin are shown. The total age range for mare volcanism is 3.8 Ga to 3.1 Ga.

Source: Data from the Basaltic Volcanism Study Project. 1981. *Basaltic Volcanism on the Terrestrial Planets.* Pergamon Press, New York. Courtesy Lunar and Planetary Science Institute.

The stratigraphic evidence that the basalts are not directly related to the excavation of the multi-ringed basins was summarized earlier. The mare basalts are therefore easier to study than the highland rocks because they are largely intact rocks with some similarities to terrestrial basalts. They occupy about 17% of the surface of the moon (30% of the near side), but only about 1% of the lunar crustal volume so that they represent a thin infilling of the multi-ringed basins (~0.5–4 kilometers thick).[7] Compositionally they are mafic basalts (25–50% plagioclase with calcic pyroxenes) and have high Ca/Al ratios compared to the highland rocks. They also have high FeO contents (~20%) and highly variable TiO_2 contents, largely represented by the mineral ilmenite ($FeTiO_3$). They are commonly classified into three groups[37]: high Ti basalts (8–14%), low Ti basalts (1.5–5%), and very low Ti basalts (<1.5%). Experimental studies indicate they represent partial melts of the lunar mantle at a depth of 150–500 kilometers, the heat for remelting presumably supplied by radioactive elements such as U, Th, and K. The impact basins provided porous conduits in the crust for the basaltic magma to reach the surface from the mantle and also provided topographic depressions in which the basalt flows ponded. The highly variable TiO_2 contents likely indicate shallow fractionation of ilmenite from a Ti-rich source and some magmas likely had titanium in their source.[37]

Rare earth elements (REE) were introduced in Chapter 8. The REE pattern of the mare basalts are characterized by negative Eu anomalies, whereas the lunar highland rocks are characterized by positive Eu anomalies (Figure 10.5).[38] The complementary relationship suggests the mare basalt were derived from a Eu-depleted source, namely the residue left behind after the lunar highlands separated by plagioclase floatation at about 4.5 Ga. Initially the magma ocean crystallized olivine and Ca-poor pyroxene. After about 70% crystallization, Ca-bearing pyroxene, plagioclase, and ilmenite were added, with the plagioclase separating by floatation to form the highland crust. The remaining crystals sank to form olivine-pyroxene-ilmenite cumulates. These rocks were also depleted in strontium and aluminum due to the separation of plagioclase, resulting in the mare basalts also being depleted in these elements. The mare basalts are therefore a product of remelting a previously fractionated moon produced during early lunar crust formation.[23] Lunar scientists, with the exception of Ringwood, were shy of illustrating their ideas with cartoon illustrations, so Figure 10.6 is the author's attempt at illustration of the magma ocean hypothesis based on descriptions in the literature.

ORIGIN OF THE MOON

The Apollo astronauts left four seismometers on the lunar surface in order to monitor lunar earthquakes and were able to derive a lunar crustal structure from the seismic results. Primary (P) wave velocities on the moon increase from about 6.8 km/sec to 8.0 km/sec at a depth of about 55–60 kilometers; this is taken as an average of the lunar crustal thickness. The crust on the far side is thicker than on the near

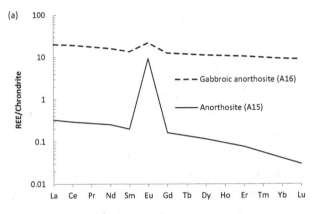

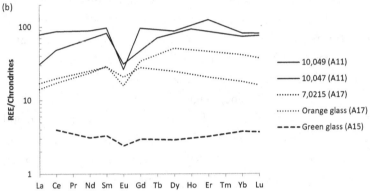

FIGURE 10.5 (a) Rare earth elements (REE) normalized to chrondrite meteorites for Apollo 15 anorthosite and Apollo 16 gabbroic anorthosite. Note the positive europium anomalies.

Source: Taylor, S. R. 1975. *Lunar Science: A post-Apollo View.* Pergamon Press, New York. Courtesy Elsevier.

(b) Rare earth elements (REE) in mare basalts. Note negative europium anomalies.

Source: Basaltic Volcanism Study Project. 1981. *Basaltic Volcanism on the Terrestrial Planets.* Pergamon Press, New York. Courtesy Lunar and Planetary Institute.

side by about 10 kilometers, possibly explaining the greater abundance of mare lavas on the nearside. The moon's mantle is divided into an upper mantle (400 to 60 kilometers) and a lower mantle (1,000 to 400 kilometers) with a possible small iron core no larger than

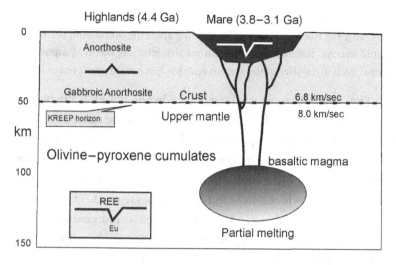

FIGURE 10.6 Formation of mare basalts by partial melting of an already fractionated upper mantle. The lunar crust formed by floatation of plagioclase-rich rocks leaving behind a residue of olivine and pyroxene (Ca-rich and Ca-poor) and ilmenite cumulates in the upper mantle. Schematic rare earth element (REE) patterns are indicated for mare basalts (see Figure 10.5b) and for the highland crust (see Figure 10.5a).

350 kilometers across. Lunar earthquakes are confined to the lower mantle and are triggered by the Earth's gravitational attraction. The magma ocean hypothesis requires a heat source, which places some constraints on lunar origin.

Traditionally, it has been suggested that the Earth's moon originated in one of three possible ways:

- The moon formed elsewhere in the solar system and was captured by the Earth. This mechanism is deemed as highly improbable; the moon would most likely bypass the Earth or collide with it rather than be captured. Furthermore, any similarities between the moon's composition and the Earth's mantle, as observed, would then be fortuitous.
- The Earth and the moon formed in the same vicinity from the same primordial cloud as a binary planetary system.
- The moon formed by fission from the Earth, an idea originally developed by George Darwin,[39] an astronomer and son of Charles Darwin. In this model, the oscillation period of a fluid Earth is only a few hours and would be

enhanced by solar tides of similar period which would cause resonance, increasing the tides and eventually splitting apart the moon from the Earth. Harold Jeffreys, the influential British geophysicist (see Box 6.1), argued against such instability in the system and this mechanism went out of favor.

The Australian geochemist, A. Edward ("Ted") Ringwood, noted that the density of the moon (3.3 g/cc) was the same as that of the Earth's mantle and proposed in his book, *Origin of the Earth and the Moon*,[40] based largely on trace element similarities between the Earth's mantle and lunar rocks, that the moon was derived from the Earth shortly after the Earth's iron core had segregated. The segregation of the core would have changed the angular momentum of the Earth and led to the splitting off of lunar material from the Earth's mantle, explaining the depletion of the moon in iron and the similarity of the moon's basaltic composition to the Earth's upper mantle.

Currently the most popular origin of the moon is a variation of the fission hypothesis in which a large impactor, the size of Mars, impacted the Earth during the terrestrial planet accretionary period (circa 4.5 Ga), and that this massive collision led to the formation of the moon.[41] This model, called the collision hypothesis, is largely based on supercomputer simulations, but is consistent with Ringwood's chemical arguments regarding the apparent similarity of the lunar composition and the Earth's upper mantle. The role of the impactor itself in the composition of the moon is still debated.

REFERENCES

1. Galilei, Galileo. 1610 (2004). *Sidereus Nuncius*. Trans., E. S. Carlos, London, 1880. P. Barker (ed.), Byzantium Press, Oklahoma City.

2. Baldwin, R. B. 1949. *The Face of the Moon*. University Chicago Press, Chicago.

3. Gilbert, G. K. 1893. The moon's face: the study of the origin of its features. *Bulletin of Philosophical Society of Washington*, v. 12, 241–292.

4. Urey, H. C. 1951. The origin and development of the Earth and other terrestrial planets. *Geochimica et Cosmochimica Acta*, v. 1, 209–277.

5. Shoemaker, E. and Hackman, R. J. 1962. Stratigraphic Basis for a Lunar Time Scale. In *The Moon*, Kopal, Z. (ed.), Academic Press, New York, 289–300.

6. Wilhelms, D. E. and McCauley, J. F. 1971. Geologic map of the near side of the moon. *Scale 1: 5,000,000*. United States Geologic Survey, Washington, DC.

7. Wilhelms, D. E. 1987. *The Geologic History of the Moon*. United States Geologic Professional paper 1348, Washington, DC.

8. Stuart-Alexander, D. E and Howard, K. A. 1970. Lunar maria and circular basins – a review. *ICARUS*, v. 12, 440–448.

9. Stöffler, D. and Ryder, G. 2001. Stratigraphic and isotope ages of lunar geologic units: chronological standard for the inner solar system. *Space Science Reviews*, v. 96, 9–54.

10. Taylor, S. R. 1975. *Lunar Science: a Post-Apollo View*. Pergamon Press, New York.

11. LSPET Apollo 11. 1969. *Science*, v. 165, 1211–1127.

12. The Moon Issue. 1970. *Science*, v. 167, 415–794.

13. LSPET Apollo 12. 1970. *Science*, v. 167, 1325–1339.

14. Schonfeld, E. 1974. The contamination of lunar highland rocks by KREEP: interpretation of mixing models. In Z. Kopal (ed.), *Proceedings of the 5th Lunar Science Conference*, v. 2, 1269–1286. Pergamon Press, New York.

15. Gromet, L. P., Hess, P. C., and Rutherford, M. J. 1981. An origin for the REE characteristics of KREEP. In Z. Kopal (ed.), *Proceedings 12th Lunar Science Conference*, v. 1, 903–913. Pergamon Press, New York.

16. LSPET Apollo 14. 1971. *Science*, v. 173, 681–693.

17. Apollo Lunar Investigation Team. 1972. Geologic setting of Apollo 15 samples. *Science*, v. 175, 407–415.

18. LSPET Apollo 15. 1972. *Science*, v. 175, 363–693.

19. LSPET Apollo 16. 1973. The Apollo 16 lunar samples: petrographic and chemical description. *Science*, v. 179, 23–34.

20. Schaeffer, O. A. and Husain, L. 1974. Chronology of lunar basin formation. In Z. Kopal (ed.), *Proceedings of the 5th Lunar Science Conference*, v. 2, 1541–1555. Pergamon Press, New York.

21. Apollo Field Geology Investigation Team: Apollo 17 landing site. 1973. Geologic explanation of Taurus-Littrow site. *Science*, v. 182, 672–680.

22. LSPET Apollo 17. 1973. Apollo 17 lunar samples: chemical and petrographic description. *Science*, v. 182, 659–672.

23. Warren, P. H. 1985. The Magma ocean concept and lunar evolution. *Annual Review Earth Planetary Science*, v. 13, 210–240.

24. Shih, C. Y., Nyquist, L. E., Dasch, E. J., et al. 1993. Ages of pristine norite clasts from lunar breccias 15445 and 15455. *Geochimica et Cosmochimica Acta*, v. 57, 915–931.

25. Tatsumoto, M and Rosholt, J. N. 1970. Age of the Moon: an isotopic study of U-Th-Pb systematic of lunar samples. *Science*, v. 167, 461–463.

26. Wood, J. A., Dickey, J. S., Marvin, U. B. and Powell, B. N. 1970. Lunar anorthosites. *Science*, v. 167, 602–604.

27. Wood, J. A., Dickey, J. S. Marvin, U. B. and Powell, B. N. 1970. Lunar anorthosites and a geophysical model of the moon. *Proceedings of the Apollo 11 Lunar Science Conference*, v. 1, 965–988.

28. Phinney, R. A., O'Keefe, J. A., Adams, J. B., et al. 1969. Implications of the Surveyor 7 results. *Journal of Geophysical Research*, v. 74, 6053–6080.

29. Walker, D. and Hays, J. F. 1977. Plagioclase floatation and lunar crust formation. *Geology*, v. 5, 425–428.

30. Taylor, S. R. and McLennan, S. M. 2009. *Planetary Crusts: Their Composition, Origin and Evolution*. Cambridge University Press, Cambridge.

31. Adler, I and Trombka, J. I. 1977. Orbital chemistry: lunar analysis from the X-ray and gamma ray remote sensing experiments. *Physics and Chemistry of Earth*, v. 10, 17–43.

32. Wänke, H., Palme, K., Baddenhausen, H., et al. 1974. Chemistry of Apollo 16 and 17 samples: bulk composition, late stage accumulation and early differentiation of the moon. *Proceedings of the 5th Lunar Science Conference*, v. 2, 1307–1334. Pergamon Press, New York.

33. Taylor, S. R. and Jakes, P. 1974. The geochemical evolution of the moon. *Proceedings of the 5th Lunar Science Conference*, v. 2, 1287–1305. Pergamon Press, New York.

34. Ganapathy, R. and Anders, E. 1974. Bulk compositions of the moon and the Earth, estimated from meteorites. *Proceedings of the 5th Lunar Science Conference*, v. 2, 1181–1206. Pergamon Press, New York.

35. Ringwood, A. E. 1976. Limits on the bulk composition of the moon. *ICARUS*, v. 28, 325–349.

36. Head, J. W. 1976. Lunar volcanism in space and time. *Reviews of Geophysics and Space Physics*, v. 14, 265–300.

37. Basaltic Volcanism Study Project. 1981. *Basaltic Volcanism on the Terrestrial Planets*. Pergamon Press, New York.

38. Haskin, L. A. 1989. Rare earth elements in lunar materials. In *Geochemistry and Mineralogy of the Rare Earth Elements*, Lipin, B. R. and McKay, G. A. (eds.). Reviews in Mineralogy, v. 21. Mineralogical Society America, Washington DC.

39. Darwin, G. H. 1880. On the secular changes in the elements of the orbit of a satellite revolving about a tidally distorted planet. *Philosophical Transactions of Royal Society of London*, v. 171, 713–891.

40. Ringwood, A. E. 1979. *Origin of the Earth and Moon*. Springer-Verlag, New York.

41. Stevenson, D. J. 1987. Origin of the moon: The collision hypothesis. *Annual Review of Earth Planetary Science*, 15, 271–315.

42. Orloff, R. 2004. *Apollo by the Numbers*. National Air and Space Agency, Washington, DC.

11 Welcome to the Anthropocene

A Man-Made Epoch?

It seems to us more than appropriate to emphasize the central role of mankind in geology and ecology by proposing to use the term "anthropocene" for the current geologic epoch.

– Paul Crutzen and Eugene Stoermer, 2000[1]

INTRODUCTION

The term anthropocene was introduced almost two decades ago by the atmospheric scientist (and Nobel laurate) Paul Crutzen and biologist Eugene Stoermer to indicate a new geological epoch caused by the increased intensity of human activity on the global environment.[1,2] It was proposed that man's activities had wrought changes on the planet equal to or greater than those of previous geological epochs, such as the Pleistocene-Holocene transition, and that the current epoch, the Holocene, had ended. The term has since gained widespread currency in both the scientific literature,[3–5] and in a variety of commentaries,[6–8] in addition to an academic journal that deals with man-made environmental changes entitled *The Anthropocene* published by Elsevier since 2013. For a signal to be used to define a new geological epoch, it must be (1) global in extent, (2) synchronous, and (3) long-lasting so that it could be recognized by future geologists. Moreover, the changes should be greater than or equal to changes upon which previous natural geological subdivisions were based.

Currently geologists have several rigorous criteria to be met before a new geological subdivision can be formalized.[9] Those skeptical of the anthropocene proposal have asked whether it is a scientific decision or a political statement,[10] and whether the idea belongs to stratigraphy or pop culture.[11] Others make the point that it is premature to define a new geologic subdivision that has barely begun and that the decision should be left up to future generations who will have the benefit of hindsight.[12] A corollary of defining a new geologic subdivision is that it should be a lithological unit that can be mapped, globally – a criterion that is difficult to satisfy in light of the current proposals.

DEFINING THE ANTHROPOCENE

The global changes attributed to the anthropocene include global warming due to increased greenhouse gas emissions and melting of the cryosphere, acidification of the oceans and interruption of the erosion and depositional sedimentary cycles due to land-use changes (e.g., deforestation) and dam-building, spread of agriculture and urbanization, and human interference in the carbon, nitrogen, and phosphorous cycles.[3] Decrease in biodiversity due to habitat loss and increase in extinction rates are also included in a possible definition of the proposed anthropocene. The original proposal by Crutzen and Stoermer was that human effects are equal to or greater than natural processes on the planet's biosphere, atmosphere, hydrosphere, and cryosphere. A working group of the Subcomission on Quaternary Stratigraphy is currently seriously considering formalizing the new epoch,[8] but fans of the anthropocene concept should not hold their breath.

Geologists don't take lightly to changes or additions to the geological timescale. If the anthropocene were to be formalized as a new geologic epoch, it would have to go through various committees in the International Commission on Stratigraphy, and then the International Union of Geological Sciences for ratification.[10] The Ordovician Period, for example, was proposed by Lapworth in 1879 to resolve the Murchinson-Sedgwick dispute over their overlapping claims of the Silurian and Cambrian systems, respectively (see Chapter 2). The Ordovician period was not officially recognized by the International Union of Geological Sciences until 1960, eight decades after the original proposal.[13] The anthropocene may never be formalized as a geologic epoch, but would likely continue to be used in an informal way by both scientists and nonscientists alike.

The development of the Phanerozoic time scale was outlined in Chapter 2. It was pointed out that the geologic timescale comprises three main components. The rock record is subdivided into chronostratigraphic units based on a relative timescale with decreasing hierarchical levels of Eonothem, Erathem, System, Series, and Stage. The base of a chronostratigraphic unit is defined by a Global Stratotype Section and Point (GSSP) or "golden spike."[13] Each GSSP should be within an interval of continuous sedimentation above and below the marker horizon (i.e., it should contain no unconformities), and it should be well exposed, easily accessible, and possible to correlate to other localities on a regional and global scale. It should be defined by as many markers as possible, for example, distinctive lithology, fossil content, chemical signature, or magnetic stratigraphy. As mentioned, it is not uncommon for agreement among the geological community to take several decades before a GSSP is agreed upon. After agreement has been reached, the final event is the placement of a golden spike to mark the exact position in the GSSP locality outcrop, sometimes together with an explanatory metal plaque.[10]

The second part of the geologic timescale requires a chronometric scale measured in years before the present (taken as 1950). This is typically done using radiometric or isotope ages (see Chapter 8), but also includes astronomical (Milankovitch) cycles and annual layers in ice cores and in the early days of geology involved an assumption of constant sedimentation rates. The decreasing hierarchical levels in the chronometric or timescale are Eon, Era, Period, Epoch, and Age. So when we speak of a sequence of rocks that belongs to the Jurassic System, the corresponding time interval is the Jurassic Period. The final part is to combine the chronostratigraphic and chronometric scales to yield a calibrated geologic timescale.[13] The chronostratigraphic scale is agreed upon by convention, but the chronometric scale is measured (e.g., isotope ratios). If the anthropocene were to be formalized, it could be at different hierarchical levels – age and epoch being the most likely possibilities. At the lowest level (age), it would then be merely a subdivision of the late Holocene epoch. Regardless of the hierarchical level chosen, a GSSP (golden spike) for the base of the anthropocene needs to be chosen and assigned a chronometric scale or age. Much of the discussion in the literature regarding the anthropocene is focused on the definition of an anthropocene series (a stratigraphic unit) and an anthropocene epoch, namely its age.[14]

The different proposals for the beginning of the anthropocene can be grouped into three temporal periods: pre-Industrial Revolution dates (also called early anthropocene proposals); a date corresponding to the Industrial Revolution (circa 1800 A.D.); and a mid-twentieth century date (circa 1950).[15] Three forces with a multiplier effect are commonly thought to be responsible for driving environmental impact by humans, namely: accelerated population growth, advances in technology, and increased consumption of resources. The question then is in which of these three time periods did these three factors come together to cause human impacts that equal those of natural processes, whether it is a change in the composition of the atmosphere or the amount of sediment moved or the amount of nitrogen produced for agriculture.[3]

As an example of an early anthropocene date, an anthropologist and archaeologist team at the Smithsonian Institution in the United States has proposed that the anthropocene coincides roughly with the Pleistocene-Holocene boundary and is represented by the domestication of plants and animals – corresponding to the beginning of agriculture.[16] This definition has the advantage that a new GSSP is not required as the Pleistocene-Holocene boundary is already defined in a Greenland ice core,[17] thereby avoiding all of the potentially lengthy formalities outlined here. However, the archaeological record reflecting the beginning of agriculture is very sparse, suggesting the human population was also sparse or that those archaeological sites have not been preserved – in either case, the preserved anthropogenic signal is weak. Moreover, agriculture began in different places at different times, spreading outward from the Levant (eastern Mediterranean) to more northern areas over a protracted period of time.[18]

A second early anthropocene proposal uses soil horizons (anthrosols) that are about 2000 years old as the base of the anthropocene, as these horizons reflect the rise of several different civilizations.[19] This proposal has not met with much acceptance, partly because soil horizons have been repeatedly reworked, are time transgressive, and are not easily preserved.[20]

A third early anthropocene proposal that has gained greater traction, judging by the large volume of literature devoted to it (see for example the Holocene special issue),[21] is the Ruddiman hypothesis. Climatologist William Ruddiman suggested that Holocene CO_2 and CH_4 variations at about eight and five thousand years ago due to land clearing by burning (CO_2 increase) and expansion of rice agriculture (CH_4 increase), respectively.[22,23] The CH_4 record for the Holocene and part of the Pleistocene was shown earlier in Figure 9.6. The record for the past 15 ka is shown here together with the solar radiation or insolation curve from 15 ka to 5 ka (Figure 11.1). The methane concentration tracks the insolation curve fairly well (except for the Younger Dryas which has an endogenic cause in the north Atlantic Ocean; see Chapter 9), but the CH_4 curve departs from the insolation curve and shows a gradual increase after ~5 ka. Because a solar radiation decrease predicts a decrease in greenhouse gases at this time, Ruddiman argues this increase has a man-made origin. A human cause for the methane increase is still a controversial idea largely because human population densities and the intensity of rice agriculture may not have been sufficient to alter the composition of the atmosphere at those times.[24] A study of archaeological sites in China and India, however, suggests a man-made cause for the methane increase between five thousand and three thousand years ago is plausible.[25] If the anthropocene concept demands that the changes observed are outside the natural envelope of

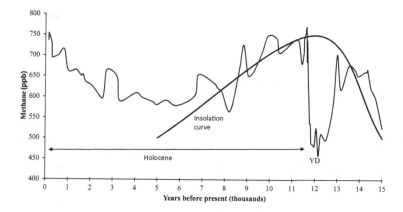

FIGURE 11.1 Methane concentration in the Greenland GISP2 ice core from 15 ka to pre-industrial times shown together with the solar insolation curve to 5 ka (heavy line). With the exception of the Younger Dryas (YD), the methane concentration and solar insolation curves are in agreement up to 5 ka at which point the methane curve begins to rise. This rise is interpreted as man-made due to the rise of wet rice agriculture about five thousand years ago.
Sources: NOAA and Ruddiman, W. F., Crucifix, M. C,. and Oldfield, F. A. 2011. Introduction to the early anthropocene. *The Holocene, spec. issue,* v. 21, 713.

observed methane variations for the Holocene, then the Ruddiman hypothesis would not meet that criterion (Figure 11.1).

The original proposal of Crutzen and Stoermer attributed the anthropocene to the Industrial Revolution (circa 1800 A.D.), corresponding to the burning of coal, the development of the steam engine, and the corresponding increase in greenhouse gases documented in ice cores at both poles at about this time. The Industrial Revolution in Britain is usually associated with a switch from a pastoral lifestyle in rural areas based on agriculture to high-density populations in industrial cities, fueled by coal and the steam engine. The main source of energy until this time was wood (with wind and water playing a lesser role), which led to widespread deforestation in Britain and forced a switch to coal around 1850. Steam engines played a major role in pumping water out of mines so that deeper and deeper coal could be mined.

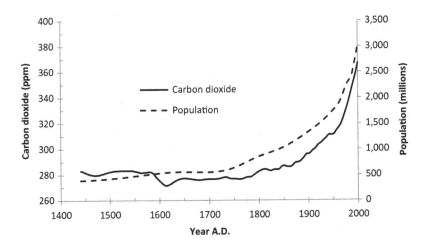

FIGURE 11.2 Global population (solid line) and CO$_2$ concentration (dashed line) from 1400 AD to today. Both curves follow the same trend. Departure from preindustrial levels begins in the 1800s, corresponding to the Industrial Revolution. Accelerated population growth and CO$_2$ concentration increase begin in the 1950s.
Sources: United Nations Population Division and NOAA.

The increase in the population around this time is due to better sanitation and expanded food production (Figure 11.2). By 1800 Earth's population had reached one billion and then doubled by 1927, and doubled again to four billion by 1974; today the global population number is headed toward 7 billion. The carbon dioxide concentration in the atmosphere, due to burning of fossil fuels (initially coal, followed by oil and natural gas) and more recently together with increased cement production, follows the exponential population increase closely (Figure 11.2). Today's carbon dioxide concentration is about 400 parts per million (first reached in 2013), from about 280 parts per million in preindustrial times, and is outside the natural range for the past 800,000 years based on the Antarctic ice cores (see Chapter 9). The three factors mentioned above for increased human environmental impact namely, population growth, advance in technology, and increased use of resources were present at this time.

BOX 11.1 **Annual CO_2 Increase in the Atmosphere**

In 2015 global fossil fuel burning and cement production (which also produces CO_2) was 36.2 billion metric tons. The major sinks for CO_2 in 2015 were the oceans (estimated at 11.1 billion metric tons) and the land (estimated at 6.9 billion metric tons). This gives net emissions to the atmosphere of 18.2 billion tons or 18.2×10^9 tons. To calculate the effect of this amount of carbon dioxide being injected into the atmosphere, we divide it by the mass of the atmosphere which is 5×10^{15} metric tons:

$$\frac{18.2 \times 10^9}{5 \times 10^{15}} = 3.64 \times 10^{-6} \text{ or 3.64 ppm.}$$

We need to make a correction to this figure of 3.64 parts per million because air and CO_2 do not have the same molecular weight. The gram molecular weight of CO_2 is 44 grams, and that of air is 29 grams. We correct the number by multiplying by 29/44, or 0.66, giving 2.4 parts per million. The average measured increase of CO_2 at Mauna Loa, Hawaii, is 2.3 parts per million per year over the past decade. The biggest uncertainty in this calculation is the size of the land and ocean sinks, which is about ±10% for each sink. By acting as a sink for CO_2, the oceans are becoming increasingly acidic, which is cause for concern for organisms such as coral reefs that secrete $CaCO_3$ for their shells.

Source of data: Carbon Dioxide Information Analysis Center (CDIAC)

It has been suggested that geochemical anomalies (major or trace elements and isotope ratios) could be used to assign an age to the base of the anthropocene,[26] and that ice cores which record these anomalies could be used as a GSSP locality[12], in a similar way that the beginning of the Holocene was defined in the Greenland NGRIP ice core.[17] An argument against using ice as a record of a geological boundary is its impermanence, especially in a warming environment. Figure 11.3 shows the levels of the toxic metals (lead and cadmium) over the period 1770–2000 AD in the southern Greenland ice core

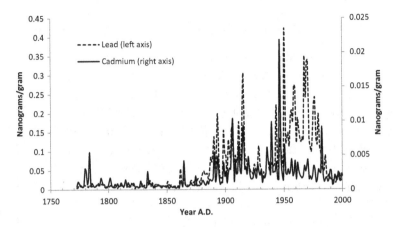

FIGURE 11.3 Concentration of lead (solid line) and cadmium (dashed line) in the ACT2 Greenland ice core. The curves can be divided into three time intervals: preindustrial levels (pre-1860), the 1900s, and the 1950s. The low levels in the 1930s corresponds to the Great Depression. The high levels of lead and cadmium in the 1900s are attributed to coal-burning in Europe and North America. The high levels in the 1950s are attributed to industrialization and the addition of lead to gasoline. Low levels today are attributed to the 1970s Clean Air Act and similar legislation in Europe.

Sources: NOAA and McConnell, J. R. and Edwards, R. 2008. Coal burning leaves toxic heavy metal legacy in the Arctic. *Proceedings Natural Academy Science,* v. 105, 12140–12144.

ACT2.[27] The record can be divided into three periods: pre-industrial values, with a sharp pulse beginning in the 1860s increasing to ten times preindustrial values in the 1900s, followed by a decrease in the 1930s during the global economic downturn. The spikes in the 1900s are attributed to increased coal-burning in Europe and North America, as there is a good correlation of these metals with black carbon in the ice core.[27] Values increase again in the 1950s, corresponding to increased burning of hydrocarbons and the introduction of lead to gasoline as an anti-knock agent in combustion engines. The Clean Air Act of the 1970s in the United States and similar legislation in Europe saw a reduction in these metals in the environment. Figure 11.2 suggests a beginning to the anthropocene at about 1850 when CO_2 and the population begin to depart from preindustrial values and a similar date is indicted by the toxic metal record (Figure 11.3).

There appears to be a growing consensus, however, that a younger date marks the base of the anthropocene at around 1950. In an unusually lengthy paper for *Science* magazine (eleven pages), Colin Waters of the British Geological Survey and twenty-two coauthors present a wide variety of evidence in favor of this younger date, as records show that at this time several parameters lie outside the natural variations seen in the Holocene.[3] Artificial radionuclides (^{90}Sr, ^{137}Cs and ^{239}Pu) show a global spike in the 1950s and early 1960s in ice cores, corresponding to testing of nuclear weapons.[26] New materials, such as plastics which appear in abundance in the 1950s and are still increasing in both the marine and terrestrial environments, are long-lived materials that will survive into the future in sedimentary deposits.[27] Concrete and aluminum are also abundant globally.[3] The cumulative production of concrete to date is sufficient to cover the entire surface of the Earth with one kilogram of concrete per square meter. Attempts to identify a stratigraphic basis for the anthropocene, however, have so far been unsuccessful.[14,20,29]

THE ANTHROPOCENE AS AN ARCHAEOLOGICAL AGE?

The original proposal for the anthropocene was to designate it as a new geological epoch. This would imply that the anthropocene series, however defined, would be a mappable unit globally, just as, for example, the Lias Series in the Jurassic System can be mapped as a global, lithologic unit. As the geologists Autin and Holbrook noted, "Presently, we are left to map a unit conceptually rather than conceptualizing a mapable stratigraphic unit."[12] Anthropogenic deposits to future workers would have more in common with archaeological sites rather than mapped geological formations. Archaeology is defined as the scientific study of the life and the culture of past peoples by means of excavation of habitations, bones, and artifacts. To future archaeologists, landfills, for example, would be similar to ancient archaeological sites, self-contained with distinct boundaries and of limited areal extent.

The archaeological chronology is more informal compared to the geologic timescale outlined in this chapter. The Danish archaeologist Christian J. Thomsen (1788–1865) identified the Stone Age, followed by the Copper, Bronze, and Iron Ages. The Stone Age was subsequently subdivided into Paleolithic, Mesolithic, and Neolithic, the latter encompassing the beginning of agriculture and pottery-making. Newer age subdivisions were added as needed in different places and times. Unlike geological boundaries, which have a unique age, the boundary between archaeological subdivisions have an age *range* reflecting the time transgressive or diachronous nature of major events in human history. Anthropogenic deposits, such as those reflecting industrialization or technological advances, are also diachronous.[14] In other words, the anthropocene has more in common with archaeology than geology. Moreover, archaeological subdivisions need not be able to be mapped over a large area, and certainly not globally. Future archaeologists might someday identify the Coal Age or the Plastics Age or the Landfill Age, and so on. The onerous, if not impossible, task of identifying a GSSP would no longer exist. I suggest we leave anthropogenic deposits and the anthropocene to archaeologists who are better equipped to excavate and classify these deposits, now and also into the future. Archaeologists appear to be willing to undertake the task.[16,19,30]

REFERENCES

1. Crutzen, P. J. and Stoermer, E. F. 2000. The Anthropocene. *Global Change Newsletter*, v. 41, 17–18.
2. Crutzen, P. J. 2002. Geology of Mankind. *Nature*, v. 415, 23–23.
3. Waters, C. N., Zalasiewicz, J., Summerhayes, C., et al. 2016. The anthropocene is functionally and stratigraphically distinct from the Holocene. *Science*, v. 351, 137–148.
4. Vince, G. 2011. An Epoch Debate. *Science*, v. 334, 32–37.
5. Crutzen, P. J. and Steefen, W. 2003. *Climatic Change*, v. 61, 251–257.
6. The Economist. A man-made world. 2011. May 28, 81–83.
7. Editorial. The Human Epoch. *Nature*, v. 473, 254.

8. Jones, N. 2011. Human influence comes of age. *Nature*, v. 473, 133.

9. Gibbard, P. L. and Walker, M. J. C. 2014. The term "anthropocene" in the context of formal geologic classification. In *A Stratigraphical Basis for the Anthropocene*, Waters, C. N., Zalasiewicz, J. A., Williams, M., Ellis, M. A., and Snelling, A. M. (eds.). Geological Society of London, special publication 395, 29–37.

10. Finney, S. C. and Edwards, L. E. 2016. The "anthropocene" epoch: Scientific decision or political statement. *GSA Today*, v. 26, 4–10.

11. Autin, W. J. and Holbrook, J. M. 2012. Is the anthropocene an issue of stratigraphy or pop culture? *GSA Today*, v. 22, 60–61.

12. Wolfe, E. W. 2014. Ice Sheets and the anthropocene. In *A Stratigraphical Basis for the Anthropocene*, Waters, C. N., Zalasiewicz, J. A., Williams, M., Ellis, M. A., and Snelling, A. M. (eds.). Geological Society of London, special publication 395, 255–263.

13. Gradstein, F., Ogg, J., and Smith, A. 2004. *A Geologic Time Scale*. Cambridge University Press, Cambridge.

14. Zalasiewicz, J., Williams, M., and Waters, C. N. 2014. Can an anthropocene series be defined and recognized? In *A Stratigraphical Basis for the Anthropocene*, Waters, C. N., Zalasiewicz, J. A., Williams, M., Ellis, M. A. and Snelling, A. M. (eds.). Geological Society of London, special publication 395, 39–53.

15. Waters, C. N., Zalasiewicz, J. A., Williams, M., Ellis, M., and Snelling, A. M. 2014. A stratigraphical basis for the anthropocene? In *A Stratigraphical Basis for the Anthropocene*, Waters, C. N., Zalasiewicz, J. A., Williams, M., Ellis, M. A., and Snelling, A. M. (eds.). Geological Society of London, special publication 395, 1–21.

16. Smith, B. D. and Zeder, M. A. 2013. The onset of the anthropocene. *Anthropocene*, v. 4, 8–13.

17. Walker, M., Johnsen, S., Rasmussen, S. O., et al. 2009. Formal definition and dating of the GSSP (Global Stratotypes Section and Point) for the base of the Holocene using the Greenland NGRIP ice core, and selected auxiliary records. *Journal of Quaternary Science*, v. 24, 3–17.

18. Bar-Yosef, O. 1998. The Natufian culture in the Levant, threshold to the origins of agriculture. *Evolutionary Anthropology*, v. 6, 159–177.

19. Certini, G. and Scalenghe, R. 2011. Anthropogenic soils are the golden spikes for the Anthropogene. *The Holocene, spec. issue*, v. 21, 1269–1274.

20. Gale, S. J. and Hoare, P. G. 2012. The stratigraphic status of the anthropocene. *The Holocene*, v. 22, 1478–1481.

21. Ruddiman, W. F., Crucifix, M. C,. and Oldfield, F. A. 2011. Introduction to the early anthropocene. *The Holocene, spec. issue*, v. 21, 713.

22. Ruddiman, W. F. 2003. The anthropogenic greenhouse era began thousands of years ago. *Climatic Change*, v. 61, 261–293.

23. Ruddiman, W. F., Ketzbach, J. E., and Vavrus, S. J. 2011. Can natural or anthropogenic explanations of late-Holocene CO_2 and CH_4 increases be falsified. *The Holocene, spec. issue*, v. 21, 865–879.

24. Crutzen, P. J. and Steffen, W. 2003. Editorial. How long have we been in the anthropocene era? *Climatic Change*, v. 61, 251–257.

25. Fuller, D. Q., Van Etten, J., Manning, K., et al. 2011. The contribution of rice agriculture and livestock pastoralism to prehistoric methane levels. *The Holocene, spec. issue*, v. 21, 743–759.

26. Galuszka, A., Migaszewski, Z. M., and Zalasiewicz, J. 2014. Assessing the anthropocene with geochemical methods. In *A Stratigraphical Basis for the Anthropocene*, Waters, C. N., Zalasiewicz, J. A., Williams, M., Ellis, M. A. and Snelling, A. M. (eds.). Geological Society of London, special publication 395, 221–238.

27. McConnell, J. R. and Edwards, R. 2008. Coal burning leaves toxic heavy metal legacy in the Arctic. *Proceedings of the National Academy of Science*, v. 105, 12140–12144.

28. Zalasiewicz, J., Waters, C. N., do Sul, J. I., et al. 2016. The geological cycle of plastics and their use as a stratigraphic indicator. *Anthropocene*, v. 13, 4–17.

29. Ford, J. R., Price, S. J., Cooper, A. H., and Waters, C. N. 2014. An assessment of lithostratigraphy for anthropogenic deposits. In *A Stratigraphical Basis for the Anthropocene*, Waters, C. N., Zalasiewicz, J. A., Williams, M., Ellis, M. A., and Snelling, A. M. (eds.). Geological Society of London, special publication 395, 55–89.

30. Edgeworth, M. 2014. The relationship between archaeological stratigraphy and artificial ground and its significance in the anthropocene. In *A Stratigraphical Basis for the Anthropocene*, Waters, C. N., Zalasiewicz, J. A., Williams, M., Ellis, M. A. and Snelling, A. M. (eds.). Geological Society of London, special publication 395, 91–108.

12 The Structure of Geological Revolutions

INTRODUCTION

Thomas Kuhn's book the *Structure of Scientific Revolutions*, published in 1962[1] and reprinted in 1970 (with a postscript),[2] is thought to be one of the most important works on the philosophy of science in the twentieth century.[3,4] Kuhn, an American historian (1922–1996) sought to show that by taking a historical view of science, which was more than just a chronology or anecdotal report, insight could be gained into how the scientific process itself worked. This was at a time after an influential group of European philosophers and their followers moved to the United States during World War II and exerted a powerful influence on academic philosophy in the United States until the 1960s. They were called the logical positivists, and while they had a high regard for the ability of the natural sciences, especially mathematics and logic, to solve important problems, they paid little attention to the history of science. This made Kuhn's book, based entirely on history, all the more remarkable.

By examining the history of the physical sciences, mainly physics and chemistry (Kuhn himself was trained as a physicist), he tried to show that science evolved through a series of stages that displayed a recurrent pattern in the evolution from immature to mature stages. He noted that immature sciences commonly had several competing hypotheses or fundamental theories (which he called paradigms), and that as time went on, one of these theories or paradigms would win out. Kuhn envisaged the science would then enter into the following stages[1]:

- Normal science for Kuhn is a puzzle-solving stage in which the daily routine of scientists is to polish the stump, so to speak, after the tree trunk had been felled (namely, after a major scientific breakthrough had already been established). It was a "mobbing-up operation"[1] to show that the new paradigm worked well by collecting new data and performing new experiments, confirming the validity of that paradigm. Alternative paradigms are not being tested at this stage.

- Anomalies begin to appear in the scientific data, and as a result sometimes new discoveries are made while the existing paradigm remains intact. Most anomalies are ignored and attributed to either poor experimentation or faulty equipment. Normal science continues.

- Anomalies begin to pile up to the point that they cannot be easily fit into the current scheme. Modifications to the paradigm are made to accommodate the new observations.

- A crisis is reached when it is acknowledged that the current paradigm is no longer acceptable.

- The response to the crisis is a scientific revolution in which a new paradigm is proposed. The old paradigm is abandoned and replaced with the new one, with new rules and new assumptions. Adherents to the old paradigm can no longer communicate with supporters of the new paradigm because their assumptions and terminology are different; they tend to talk past one another (Kuhn called this incommensurability). This then is followed by normal science again under the new paradigm.

In addition to these stages in the development of the scientific process, which were novel at the time, Kuhn presented several much more controversial ideas that contradicted the logical positivists. The traditional view in the philosophy of science, and for the practicing scientist, was that a choice could be unequivocally made between any two theories by seeing which theory best fit the data or observations. Contradicting this, Kuhn maintained that data was "theory-laden" – that data did not speak for itself, but rather was dependant on the theory through which the observer viewed it. Kuhn thought that scientists aligned with different theories saw the same facts differently. If true, this had profound and radical implications, namely that there are no objective facts and that science itself is not a rational activity. In his postscript to the 1970 edition of *Structure* (as he referred to his book), he replied to some of the criticisms of his earlier edition and backed off most of these more radical ideas.[4]

Two of Kuhn's chief examples of a paradigm switch are the transitions from Ptolemic (Egyptian) astronomy to that of Copernicus, and the transition from Newtonian physics to Einsteinian physics. All five of the stages outlined in Kuhn's *Structure* are present in the history of geology as outlined in the previous chapters of this book, so it is interesting to evaluate how well, or poorly, that template applies to geology. In his 1974 "Second Thoughts on Paradigms," Kuhn allows that the pattern outlined above also applies equally well to the pre-paradigm phase.[5] In other words, it applies to an immature science, such as geology, prior to Lyell. The philosopher of science, Ian Hacking, emphasizes this same point.[2] The main conclusion of this chapter is that Kuhn's ideas apply quite well to the history of geology.[1,5]

CATASTROPHISM VERSUS LYELL'S UNIFORMITY PRINCIPLE

In the nineteenth century, geology was still in its cradle and not yet a mature science, still trying to divest itself from biblical accounts and unsure how it should proceed. Adam Sedgwick, in his presidential address to the Geological Society of London in 1830, laid down the standard for this fledgling science, which is worth quoting again from Chapter 1: "We must banish *a priori* reasoning from the threshold of our argument and the language of theory can never fall from our lips with any grace or fitness, unless it appear as the simple enumeration of those general facts, with which by observation alone, we have at length become acquainted."[6] Sedgwick's reference to *a priori* reasoning no doubt refers to the subtitle of Lyell's *Principles* (Chapter 1): "Being an attempt to explain the former changes of the Earth's surface by reference to causes now in operation."[7] In his 1829 letter to Murchinson, Lyell elaborates (Box 1.1): "that no causes whatever have from the earliest time to which we can look back, to the present, ever acted, but those now acting; and that they never acted with different degrees of energy from that which they now exert."[8] This statement expresses a new paradigm in that geologic processes can be understood by looking solely at processes operative today; this is a clear break with the catastrophists' view of geological history, which was marked by Earth revolutions. In *Structure*, Kuhn defined two characteristics of a new paradigm[1]:

- Sufficiently unprecedented achievements to attract an enduring group of adherents, and
- Sufficiently open-ended, with the potential for the adherents to solve lots of new problems.

As Lyell's *Principles* went through twelve editions until his death, and was a very popular book, we can say that Lyell had an enduring group of adherents. And his new way of looking at geological processes had the potential to address a wide variety of geological problems. A brief look at the topics that Lyell addressed in *Principles* indicates that he covered most geologic phenomena known at that time in the historical period. In this regard, Lyell's uniformity principle satisfies the two criteria Kuhn defined, and his views therefore represent a new paradigm.

How did Lyell know that past geologic processes "never acted with different degrees of energy."[8] The answer is that he did not know, because the data needed to make such an assertion did not exist at the time. It was indeed an *a priori* assumption, as Sedgwick claimed. How he arrived at this conclusion, on his own, is certainly worthy of further study. When examining the nineteenth-century catastrophism versus uniformity controversy in geology, as outlined in Chapter 1, the historical record favors a view that is closer to Kuhn's ideas rather than the traditional view of the logical positivists.

In a letter to Lyell in 1841, William Conybeare summarized the main points against Lyell's uniformity position (see Chapter 1). He agrees with the position of the uniformity of the laws of nature and general physical causes. But he argues that "different conditions at different times materially modified their intensity."[9] Citing the fossil record sequence from younger to older sediments – man (recent strata), mammalia (tertiary strata), reptiles (secondary strata), and fish (primary strata) – Conybeare concluded "an arrangement of progressive organization" is apparent.[9] Referring to igneous and metamorphic rocks, he noted that the older rocks display greater igneous activity compared to younger ones, and similarly that older rocks such as Silurian and Carboniferous rocks of England showed more contortions compared to younger ones. Lyell's response in earlier editions of *Principles* was that older rocks had more time to endure these deformations. Conybeare counters in his letter that these deformed rocks are unconformably overlain by horizontal and unmetamorphosed strata. Conybeare also argues that older rocks contain more volcanic activity compared to secondary and tertiary rocks. He concluded: "I can discern nothing like this regularly recurring series of uniform events."[9] In short, the facts supported the catastrophic position (in addition to Cuvier's extinctions), but Lyell had little solid evidence for his position, citing only an incomplete stratigraphic and fossil record.

Why then did Lyell's approach win out over the catastrophist position? The popularity of *Principles* may be one reason. During the Enlightenment the move away from scriptural geology and biblical literalism, with which some of the catastrophists were associated, may be another reason. Lyell's theory was simple and easily understood, and it also had the potential to explain a large number of ancient geological phenomena. Kuhn states (in Section 12 of *Structure*) that the decision to adopt one paradigm over another can only be made on faith, and that it is a conversion experience that cannot be forced.[1] He also suggests that peer pressure from other scientists may play a role. The result of the catastrophism versus uniformity debate may be surprising to the logical positivists, but was not so surprising to Kuhn. Nevertheless, the distinction between catastrophism and the uniformity position was beginning to blur by the 1850s.[10]

The rise and fall of Neptunism, which took place over a fifty-year period (1775–1825), offers a contrast to the catastrophism-uniformitarian debate above.[11] Here the two theories (Neptunism and Plutonism) were compared to the observations made on the rocks, and Neptunism was found wanting. As Werner's students went abroad to apply the theories of their Saxon professor to the outside world, they found Neptunism did not apply to other regions. Hutton had shown in 1794 that granites in Scotland were intrusive into the surrounding country rock, which contradicted Werner's idea that granites were the oldest rocks (Chapter 1). It also became clear that basaltic sills had textures and minerals similar to modern lava flows, and were therefore also igneous in origin. This controversy followed more along the line expected by the logical positivists, rather than those of Thomas Kuhn. But it is consistent with his view that an immature science commonly had more than one paradigm to contend with.

NORMAL SCIENCE: THE GEOLOGIC STRATIGRAPHIC COLUMN

Assigning the building of the Phanerozoic geologic column to normal science may seem to stretch Kuhn's ideas on that phase in scientific development too far, simply because it was a pioneering undertaking in the development of geology as a science. Table 2.2 shows that it was accomplished by a group of about ten geologists or "gentleman specialists" over the first half of the nineteenth century (the naming of the Ordovician period is an outlier in terms of timing; see Chapter 2 for details) who worked largely independently of each other, but were nevertheless part of a community with shared goals and geologic principles. They belonged to the same geological societies (in London, Edinburgh, and Paris), and published in the same scientific journals of the day. They commonly communicated with one other by correspondence.

The principles of stratigraphy are fairly rudimentary, and were established early on (Chapter 2), so that in Kuhn's sense of normal science, the paradigm did not change over the course of this enterprise. They were in the puzzle-solving stage. New discoveries were made of course – new fossils were found and new lithological associations were discovered. Rocks previously thought to belong to the

Carboniferous System were reassigned on the basis of fossils to a new geological system, namely the Devonian. Rocks of the same age but of different lithology in different locations introduced the concept of sedimentary facies, implying different environmental conditions at different geographic locations at the same time. William Smith in England recognized that rocks of different age had their own distinctive fossil assemblages, an important early discovery. As more complex (folded, faulted, and metamorphosed) rock formations of the lower Paleozoic were studied in Wales, for example, the same stratigraphic principles applied, although in a more cryptic fashion. These geologists recognized that younger rocks occurred in the cores of synclines, older rocks in the cores of anticlines. But overall, they were doing normal science – their basic paradigm did not change over the course of establishing the Phanerozoic geologic column. Lyell himself was involved in discovering part of the Cenozoic record at this time (Table 2.2). They disagreed and argued with one another, of course, but by this time the Neptunist–Plutonist debate was largely over; they were not testing different paradigms. This small group of early geologists fits very well into Kuhn's idea of a scientific community operating under the same paradigm as detailed in "Second thoughts on paradigms."[5]

CRISIS IN TECTONICS

This crisis lasted about seventy years (1890s–1960s), and it ended with the plate tectonic revolution (Chapter 7). It can be argued that the theory of Continental Drift (Chapter 6) was the first (rejected) response to the crisis in tectonics, and that plate tectonics was the second response, which today is the dominant paradigm throughout the Earth sciences (Figure 12.1). The crisis (Chapter 5) centered on the failure of the geologic community to explain several large-scale geologic phenomena on Earth: the origin of mountain belts, the correlation of similar fossils and similar geologic formations on disparate continents, the origin of geosynclines, and the nature of the ocean basins. Normal science continued in other subdisciplines (e.g., stratigraphy, paleontology, etc.).

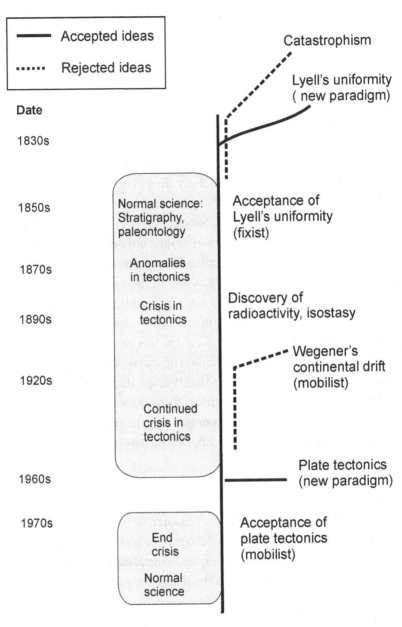

FIGURE 12.1 A timeline for the structure of geological revolutions. Early on, several hypotheses were in competition with each other. By the 1830s, Lyell's uniformity principle became the dominant paradigm, although some still adhered to catastrophism, particularly clergymen. The Phanerozoic geologic column was largely completed under "normal science" conditions. In the subdiscipline of tectonics, anomalies were mounting by the 1890s and reached a crisis stage shortly thereafter. The first response to this crisis was the theory of Continental Drift, which was largely rejected. The second response to the crisis was the plate tectonics revolution, which returned Earth science to normal science again. Dashed lines represent those ideas that did not survive.

The discovery of radioactivity in 1896 had several important implications for geology – it invalidated Kelvin's age of the Earth which was calculated assuming the Earth was slowly cooling down since its formation, without an internal heat source (Chapter 3).[12] Similarly, it also undermined the basic assumption of the contraction theory of mountain belt formation as due to the long-term cooling of the Earth. The contraction (cooling) theory for the origin of mountain chains was the dominant theory throughout the nineteenth century. In essence, this theory maintained that the cooling of the Earth caused contraction of the Earth's radius, thereby causing compression and folding of the outer crust. The theory, in a qualitative form, dates back to the time of René Descartes (Figure 5.1), but attempts to quantify it were not made until the 1870s and 1880s. For example, a study by the Rev. Osmond. Fisher concluded the radial contraction of the Earth was insufficient to cause the elevation of mountains,[13] but others concluded the amount of contraction was adequate.[14,15] None of these quantitative calculations, however, included the effects of radioactivity and therefore were of dubious validity. Harold Jeffreys, the English geophysicist, did include the effects of radioactivity and concluded, nevertheless, that contraction theory was sufficient to raise mountain belts; he maintained this opinion up until 1976 (*The Earth*, sixth edition). Parenthetically, it was recognized early on that radioactivity in the Earth was largely concentrated in the continental crust and did not affect the overall thermal history of the deep Earth.

Geological cracks, in addition to the quantitative problems mentioned, were also appearing in the contraction theory. Clarence Dutton had pointed out that mountain belts are asymmetrical, but the contraction theory did not predict such asymmetry.[16] He also said if the crust was floating on a liquid substratum, as many believed, the orientation of mountain belts should be in all directions, which was clearly not the case. Dutton wrote, "[T]his hypothesis is nothing but a delusion and a snare, and the quicker it is thrown aside and abandoned the better it will be for geological science."[17] James Dana tried to explain the mountain belt asymmetry by modifying the contraction theory.[18] He argued that because the ocean crust is stronger than continental crust, during cooling a greater force would come from the oceanic side of the mountain belt. This would explain mountain belts in North America on the east and west coasts where mountain belts verge toward the continental interior. As late as 1929, Harold Jeffreys supported this idea in his book *The Earth* (second edition).

An alternative to the contraction theory was sought in isostasy. If horizontal compression, from whatever source, caused the crust to double in thickness, then isostatic equilibrium would cause the crust to be elevated, by several kilometers. This idea gained some support among geologists at the time, including Dutton. Dutton, himself, introduced the term isostasy into the geologic literature in 1889, but the original idea is due to George Airy in 1855 (Chapter 5).

If isostasy helped solve some problems, it also raised other more serious ones. As early as the 1870s, geologists recognized similarities of fossils and sedimentary formations on different continents in the southern hemisphere such as Africa, South America, and India (Chapter 6). Edward Suess in 1904 gave the name "Gondwána Land" to this former continent. It was assumed that the ocean basins between these continents represented foundered continents. Similarly, Bertrand's tectonic correlations across the North Atlantic assumed the Atlantic basin was a foundered continent (Chapter 5). But isostatic equilibrium precluded the foundering of continents. The explanation for the similarities in fossils between these continents then became narrow land bridges of unknown origin. However, if these bridges were continental in character, they also could not have foundered.

James Hall's geosyncline concept of 1883 was criticized by James Dana as a theory of mountains with "the mountains left out" (Chapter 5).[18] Understanding how geosynclines became the locus of igneous activity and compression, and were then elevated into mountain belts would not be explained until the John Dewey and John Bird paper in 1970, approximately one hundred years later (Chapter 7). An additional tectonic puzzle at the time was understanding the emplacement mechanism of Alpine nappes and thrust sheets with large displacements.

It was clear that anomalies and unanswered questions in tectonics were beginning to pile up. Contraction theory did not explain the asymmetry common to most mountain belts, unless Dana's modification was accepted. In any case, heating from radioactivity rendered the secular cooling model invalid. It also rendered invalid the quantitative calculations on the amount of mountain belt shortening it could cause. Dutton had argued the contraction theory should be abandoned. Foundering of the continents into ocean basins was no longer a tenable idea and land bridges were pure speculation. How then to explain Marcel Bertrand's tectonic correlations across the Atlantic or similarities in geology and fossils between the Gondwanaland continents. The evolution of geosynclinal basins was also not understood. The inability to explain several of the Earth's most significant geologic phenomena indicates a crisis stage had been reached by the 1890s. This timeframe corresponds to the recognition of the importance of radioactivity and isostasy among geologist – both of these issues underscored the crisis in tectonics.

FIRST RESPONSE TO CRISIS: CONTINENTAL DRIFT

The next stage in Kuhn's model after crisis is a new theory, namely a paradigm change. In 1912 it came from an unexpected quarter, an outsider, a man trained as a meteorologist – Alfred Wegener. Continental Drift – or horizontal displacement theory, as he called it – sought to explain several of the anomalies discussed in this chapter at once, in addition to other anomalies. Kuhn notes in *Structure* that new theories often come either from younger scientists (Wegener was thirty-two when he first published his theory), or from scientists new to the field who have not been exposed to the traditional methodologies and ideas of the discipline.

Continental Drift proposed a mobilist view of global tectonics in which the continents moved thousands of kilometers in contrast to the previous fixist view. Wegener sought to explain mountain belts as due to the movement of continents through the mafic substratum, causing crumpling at the leading edge of the continent. His mobilist view would explain the tectonic, fossil, and sedimentary record connections between the elements that made up Gondwanaland. Continental Drift also explained the jigsaw-puzzle fit of the continental outlines. This new theory had only a handful of adherents and was largely rejected, but nevertheless was widely debated until the 1960s (Chapter 5). The theory of Continental Drift became widely known circa 1925 in several different languages, so the entire debate lasted about thirty-five years. The reasons for the rejection are complex, especially in the United States, and the topic is addressed more in depth in Naomi Oreskes' book.[19]

SECOND RESPONSE TO CRISIS: PLATE TECTONICS

A second new paradigm emerged over a very short period of time (1960s), namely plate tectonics, which was widely accepted by most Earth scientists, also over a very short period of time (Chapter 7). Plate tectonics corresponds well to a scientific revolution, or paradigm change, in Kuhn's terminology. However, the rapidity of its acceptance must surely have been partly due to Continental Drift and continental paleomagnetic results before it, which prepared scientific minds for such an all-encompassing paradigm shift. Overall, Kuhn's *Structure of Scientific Revolutions* fits the history of the development of geology quite well (Figure 12.1). However, some philosophers of science may have problems with this interpretation.[20,21]

REFERENCES

1. Kuhn, T. 1962. *The Structure of Scientific Revolutions*. University of Chicago Press, Chicago.
2. Hacking, I. 1970. Introductory essay. In *The Structure of Scientific Revolutions*, 2nd ed., Kuhn, T. (ed.) University of Chicago Press, Chicago.
3. Hoyningen-Huene, P. 1993. *Reconstructing Scientific Revolutions*: Thomas Kuhn's philosophy of science. University of Chicago Press, Chicago.
4. Okasha, S. 2002. *Philosophy of Science*. Oxford University Press, Oxford.
5. Kuhn, T. 1974. Second thoughts on paradigms. In *The Structure of Scientific Theories*, Suppe, F. (ed.). University of Illinois Press, Urbana, Illinois.

6. Sedgwick, A. 1830. Annual address. *Proceedings Geological Society London*, v. 1, 187–212.

7. Lyell, C. 1830–1833 (1997). *Principles of Geology* (3 vols.). Penguin, London.

8. Lyell, K. M. (ed.). 1881. *Life, Letters and Journals of Charles Lyell*. J. Murray, London.

9. Rudwick, M. 1967. A critique of uniformitarian geology: a letter from W. D. Conybeare to Charles Lyell, 1841. *Proceedings of the American Philosophical Society*, v. 111, 272–287.

10. Wilson, L. G. 1980. Geology on the eve of Charles Lyell's first visit to America, 1841. *Proceedings of the American Philosophical Society*, v. 124, 168–202.

11. Adams, F. D. 1938. *The Birth and Development of the Geological Sciences*. Williams & Wilkins Co, Baltimore.

12. Joly, J., 1911. *Radioactivity and Geology*: An account of the influence of radioactive energy on terrestrial history. Constable & Co, London.

13. Fisher, O. 1888. On the mean height of the surface elevations and other quantitative results of the contraction of a solid globe through cooling. *Philosophical Magazine*, v. 25, 7–20.

14. Davison, C. 1887. On the distribution of strain in the Earth's crust resulting from secular cooling: with special reference to the growth of continents and the formation of mountain chains. *Philosophical Transactions Society of London, A*, v. 178, 231–242.

15. Darwin, G. 1887. Note on Mr. Davison's paper on the straining of the Earth's crust in cooling. *Philosophical Transactions Society of London, A*, v. 178, 242–249.

16. Dutton, C. E. 1874. A criticism upon the contractional hypothesis. *American Journal of Science, 3rd series*, v. 8, 113–123.

17. Dutton, C. E. 1882. *Physics of the Earth*. A review. O. Fisher (ed.) *American Journal of Science, 3rd series*, v. 23, 283–290.

18. Dana, J. 1873. On some results of the Earth's contraction from cooling, including a discussion of the origin of mountains and the nature of the Earth's interior. Part I. *American Journal of Science, 3rd ser.*, v. 5, 423–443.

19. Oreskes, N. 1999. *The Rejection of Continental Drift*. Oxford University Press, Oxford.

20. Frankel, H. 1978. The non-Kuhnian nature of the recent revolution in the earth Sciences. *Proceedings of the Biennial Meeting of the Philosophy of Science Association*, v. 2, 197–214.

21. Laudan, R. 1978. The recent revolution in geology and Kuhn's theory of scientific change. *Proceedings of the Biennial Meeting of the Philosophy of Science Association*, v. 2, 227–239.

Index